# THE MALAY EXPERIMENT

The Colonial Origins and Homegrown Heroics of
The Malay Regiment

Stuart Lloyd

CAT
MAT
DOG

CatMatDog
Storytelling

Published by: CatMatDog Storytelling, Australia. ABN: 85312594003

Contact the author: stulloyd.worldsmith@gmail.com



Cover Design and Interior Photo Pages: Stuart Lloyd.
Interior typesetting: Shahid Aziz.

Available in paperback, eBook, and audiobook formats.
Paperback ISBN: 9780645328097
Malay translation version available.

# Contents

**Experiment:** "A test done in order to learn something or to discover if something works or is true."
Cambridge Dictionary.

# Foreword

I must congratulate the author Stuart Lloyd for this non-fictional military history book, *The Malay Experiment: The Colonial Origins and Homegrown Heroics of the Malay Regiment.*

An accomplished author who has painstakingly also penned several well-acclaimed books relating to the military, Stuart Lloyd surprisingly is not even a soldier. Indeed, here is a writer driven by passion for the deeds of brave soldiers who laid down their lives, made immense sacrifices, and stayed steadfast to their calling.

For him to capture the history of the military experience here in Malaysia (Malaya then), it should be a great discovery to know that he is not even a Malaysian.

This book impresses me beyond words and I humbly acknowledge that what Stuart Lloyd has contributed through this book makes up for what many of us who served in the Malay Regiment (now designated Royal Malay Regiment) — including the Malaysian Army — could not fulfil.

This book will remain a voluminous contribution making up for the absence of a dedicated military historian or writer to record all activities of the regiment since its establishment in 1933, providing a permanent reference for future generations.

In its 92 years of existence, the Malay Regiment has performed numerous services for the nation, notable among them being the 1st Malayan Emergency (1948-60); the Indonesian-Malaysia Confrontation (1962-66); the 2nd Malaysian Insurgency (1968-89); the Sarawak Communist Insurgency (1962-90) and having involvement in over 39 UN Peacekeeping Operations.

As a young officer of the 6th Battalion Royal Malay Regiment (to which I was posted upon being commissioned as an officer in 1966), I recall how we were constantly reminded of the successes of the battalion against the communist terrorists during the period of the 1st Malayan Emergency. We were so dutifully reminded that the battalion had British officers from the 1st Devonshire and Dorset Regiment seconded to the battalion during the formative stage. The first Commanding Officer of the battalion was Lt-Col OGW White.

Interestingly, the book makes mention of Lt-Col Ibrahim Ismail (who served as the Chief of Defence Force from 1970-77) as the first Malay Officer to be appointed as the Commanding Officer of the 6th Battalion Royal Malay Regiment.

In combing through Stuart Lloyd's documentation of the early Malay Regiment, I recall some of the early traditions he mentioned that were still practiced in the 1960s. One such practice was the Friday mosque parades with soldiers dressed in their traditional Malay attire — *Baju Melayu* — marching to the nearby mosque. Unfortunately, this tradition is no longer in practice today. Another interesting observation by the author is the mention of Lt Adnan Saidi who was killed in the Battle of Pasir Panjang on the island of Singapore, battling against the Japanese during World War II in 1942. In fact there is a school named after him at his village of birth namely, Sekolah Kebangsaan Lt Adnan (Lt Adnan National School) at Sungai Ramal, Kajang, in the state of Selangor.

There are many other poignant historical recollections of the Malay Regiment that may have been lost among the generation of the officers and men of the regiment today.

I believe Stuart Lloyd's latest book *The Malay Experiment* will rekindle the minds of officers of the Malay Regiment. More importantly the passionate works of Stuart Lloyd adds to the richness of the 92-year-old history and traditions that will remain etched in the annals of the regiment.

This book is a must-read for all officers of the Malay Regiment.

**Brigadier General Dato Mohamed Arshad Raji (Retired)**

**January 2025**

# Preface

Much of the action in this book — and much of the homegrown heroics that created the legend of the Malay Regiment — took place in Pasir Panjang, Singapore.

Coincidentally when I first went to work in Singapore in the late 1980s, my boss was helping me to find accommodation. 'Most of the expats live on the east coast,' she said. 'Good, I'll go and look on the west coast then,' was my immediate reply. I guess I didn't see myself living in Asia just to meet other Europeans, or *matsallehs* (mad sailors) as they're called in Malay. Rather, my goal was more to immerse myself in local culture where possible. Or maybe I am just contrarian?

I soon moved into a condo on Pasir Panjang Road, which was considered *ulu* (far out) by my work colleagues. But I loved it because it was then still bucolic: overlooking Malay *kampongs* (traditional villages), roosters crowing, and excellent food stalls serving the tastiest satays and *nasi padang*.

Just down the road from me was a massive concrete pillbox, tilted at an obtuse angle on the roadside just before Pasir Panjang Village. This piqued my curiosity, because it looked so out of place on this languid tropical island. And behind this rose the undulations of Pasir Panjang Ridge, which provided some great serpentine roads for motorcycling when visiting friends in Ulu Pandan and Holland Village. Often this would take in the stretch known as Reformatory Road. And my favourite duck rice stall was in Buena Vista, not far from home.

Little did I know then that these were all key names that featured in the Battle for Singapore that raged down the west coast of Singapore in February 1942. And of course I then knew nothing of the regiment which made its name in those same hills that month: the Malay Regiment. It would only be nearly 20 years later when I wrote my first military history book, *The Missing Years*, that I could join all of those dots, and realised that I had lived among such history and heroism in the 10 wonderful years I called Singapore home.

A further fillip came when I was asked to develop and run WW2-themed tours through Singapore, Malaysia, and Thailand on the back of that book (which was about a British rubber planter's experiences fighting in Malaya, being hospitalised at the Alexandra Military Hospital — also on the west coast of Singapore — and his hardships on the Thai-Burma Death Railway).

This involved extensive recce visits to the battlefields of Johor, plus all the wonderful coastline right up to the Thailand border. Where some of the first and fiercest battles against the Japanese were fought.

The highlight of my Malaysian travels (for business or pleasure) has always been the genial, gentle, and generous nature of the people.

Notable among these were Johari Shafie and wife Jenny who founded the Batu Muang Fort attraction (now Penang War Museum). Johari was kind enough to entertain my hair-brained scheme to borrow one of his vintage Japanese bicycles and follow the Japanese' tracks from their landing sites in Kota Bahru in the north-east all the way down to Singapore. (My knees still thank me for not following up on that idea!)

Another charming and cultured gent I've been lucky to deal with is Dato' Haji Azizan Noordin, when he was Acting Director-General of Tourism Malaysia. I regret that I was unable to take up his kind invitation to attend the MotoGP races at the time.

Another good-hearted and good-humoured Malaysian I met along the way — via that other Malayan/Malaysian institution the Hash House Harriers — is Major (Ret.) Hardy Dhaliwal, who once treated me at his home to one of the best home-cooked curries I've ever tasted, and subsequently facilitated an introduction to Brig-Gen Dato Mohamed Arshad Raji (Retired), a Malay Regiment stalwart.

And Liz Dass, then PR Manager of the E&O Hotel, who introduced me to the tropical taste sensation of Penang Laksa. During this meal she told me of her Uncle Charles who was taken off the streets and press-ganged into working as a native *romusha* labourer on the Thai-Burma Death Railway. Amazingly he managed to escape and made his way back to Malaya, traumatised by the experience. Equally incredibly, her family boasts four members of the military who have put in a combined 110 years of service for the nation (one brother was a Brigadier General, the other a Maritime First Admiral). Her father was in the Intelligence Corp during the Emergency.

Then there are those countless casual encounters of the everyday, in the local markets and hawker centres, the heart of Malaysian life. Always so warm and engaging (despite my woeful attempts at the lingo). No wonder there were questions over whether the Malay people could be martially effective.

By 2002 the Reflections at Bukit Chandu experience centre had been established on top of the hills near Kent Ridge, where the Malay Regiment's showpiece battle is commemorated. This previously little-known episode was superbly curated, and the story wonderfully told. Which further piqued my interest and curiosity, especially the derring-do of the Malay and English officers and men against the rampant Japanese. This centre soon became one of the most popular stops for my tour groups.

As a storyteller, my antenna is always up for great demonstrations of human spirit. And this had all the ingredients in spades. With its colourful colonial origins, the testing of a hypothesis in the heat of battle, and the subsequent delivery on that potential. All set against the backdrop of waning empire, and a nascent sovereign nation.

By definition, the focus is up-weighted on the period from the regiment's formation in 1933 until 1947 (when it became Royal Malay Regiment), my idea being to put across an origin story which created the template and values-system for the regiment.

I am flattered and humbled by the generous words of one of the Malay Regiment's very own leaders, Brigadier General Dato Mohamed Arshad Raji (Retired), who kindly penned the Foreword to this book. His imprimatur implies that my interpretation of the events — and especially the cultural spirit of the regiment — is close to the mark. Despite the fact that I am neither soldier nor Malaysian, as he points out. But I have lived 25 years around Asia. And in my corner, I also have a great-grandfather who was in the Imperial Light Horse who was one of just six officers who rode in to relieve the Siege of Mafeking in the Boer War. Plus my father-in-law is a retired three-star General in the Royal Thai Army. Can one gain credentials by such osmosis? Probably not, though I have certainly learned deep humility from my father-in-law and his daughter.

But while I have a bit of military understanding somewhere deep in my DNA, I frankly would not have the guts to withstand the chaotic action of a single battlefield, let alone a whole war. Which is why I find the gallant feats of the once-ordinary young men in this story so impressive and compelling.

These days the west coast of Singapore is unrecognisably built up, with high-rise condominiums cheek-by-jowl along Pasir Panjang Road (and in Pepys Road, condos don't go for less than SGD$8 million these days). Ironically Pasir Panjang is especially popular with Japanese families, who constitute the largest expat community of Singapore (as they do in many South East Asian countries). But that roadside pillbox still sits there, silent yet sinister. And now whenever I pass that area, one glimpse of it is enough to conjure up all the vicious and spirited action that took place around the area in 1942, with the Malay Regiment holding their own in the thick of it.

This then is their inspiring story. And it is my privilege to bring this story to a wider audience, to get to know better the exploits of this proud home-grown regiment. I sincerely hope I've done this story justice.

**Stuart Lloyd**
**February, 2025**

# What If?

Port Dickson, Malaya, 1933. The straight-nosed, wavy haired officer with a crescent-shaped scar on his right cheek looked every inch – and his height was considerable – the war hero. Which indeed he was.

Captain George McIllree Stanton Bruce was only 36 but had already lived two lifetimes of adventure, having signed on to the Lincolnshire Regiment as a slender 17-year-old fresh out of the all-boys Haileybury boarding school in Hertfordshire, via Sandhurst. Haileybury boasted 17 VC holders among its alumni, and the scar on Bruce's cheek was a daily reminder of what many considered to be the worst day in the history of the British Army to that point: 1 July 1915.

He'd been shot in the face during the Battle of the Somme.

After two months' medical recuperation, (Acting) Captain Bruce was back to lead his men. Next came the Battle for Cambrai in Gonnelieu, France. His company was strongly counter-attacked yet he darted from post to post cheering on his men while under heavy fire, and the enemy was finally beaten back. The 20-year-old was awarded the Military Cross for that conspicuous gallantry.

Then just a few months later the Battle for Passchendaele erupted, in the Belgian town of Ypres. While walking point for a battalion of 620 men, advancing slowly, Bruce was struck in the chest by machine-gun fire. A stretcher party was called up. As he was being carried back, a shell exploded nearby further injuring him in the leg and shoulder. He was left in the open for 12 hours, unable to be retrieved as shells thudded into the flat fields around him.

So he was just the man needed when on 1 March 1933 an experimental company of the Malay Regiment was formed with the idea of developing a Malay regiment. The idea was for an all-Malay regiment to take some responsibility for security of the Federated Malay States, but the suitability of the Malay temperament for military discipline and bravery was as yet unknown. Sure, they knew war — after all, states such as Negri Sembilan, Pahang, Selangor and Perak had been feuding since anyone could remember — and being on a peninsula, they certainly had a good pedigree in piracy.

Yet there was a military precedent, lost in the mists of Whitehall, London, apparently.

We need to sail back to 1799, to the tiny island of St Helena — in the middle of the Atlantic Ocean — which was a popular resupply port on the long and arduous journey from Europe to Asia. A number of Malays were serving on enemy ships ... possibly French or Spanish, because they'd joined forces to challenge British naval superiority, or Dutch because the Brits had taken this island from them just four years earlier. (In doing so, there were already some Malays among the 300 taken prisoner on the island.)

Now the crew, including the Malays, were taken prisoner, and made to switch sides ... but not before taking an oath of fidelity, and joining the British forces as artillerymen.

They were to be part of the British Colonial Auxiliary Forces, which mostly used local inhabitants, and were used to maintain security and order in Britain's many far-flung colonial outposts.

One of these was Bencoolen (now Bengkulu) in Sumatra. The Brits held the Padang area of western Sumatra during the Napoleonic Wars, while Holland controlled most of the rest of the Dutch East Indies (now Indonesia). In the Anglo-Dutch Treaty of 1824, Britain ceded Padang to the Dutch in exchange for control of Malacca and Singapore.

In 1801, the Malays were shortly shipped off to Ceylon (now Sri Lanka), where the Ceylon Rifle Regiment was taking shape. With these additional Malays from St Helena, the regiment was now ten companies strong, and was renamed as the Malay Regiment for the next six years, before re-designation as the 1 Ceylon Regiment, which became a light infantry unit. It was now a blend of Malays, Sinhalese, and other ethnicities, under British officers.

They acquitted themselves well in the 1817 Uva Rebellion, which was an anti-colonial uprising by Kandyan aristocracy and rebels loyal to them. Fighting mainly in the central highlands of the island, the Kandyans employed guerrilla tactics, using the dense jungle terrain to their advantage. The British responded with brutal counterinsurgency measures, including scorched-earth policies, destruction of villages, mass executions, and even deportations to quell the rebellion. It was put down within a year. Britain took over complete control of Ceylon and the regiment was disbanded some decades later.

According to an illustrated booklet that would be issued by the Colonial Government's Ministry of Information (in Malay and English) in 1941, they had proved to be "excellent soldiers" in that circumstance well over a century before.

But, could Malays form an effective *modern* fighting force?

After all, they surely could and should play some part in the defence of their own homeland, other than as volunteers in the Malay States Volunteer Rifles. And the British Indian Army was finding the cost of having the Burma Rifles garrisoned in Taiping since 1919 to be a cost-ineffective way of ensuring security.

The question about using Malays had lingered so long — initially mooted in 1902 by the royal family and sultan of Perak, sultan of Negri Sembilan, and *undang luak* (chief) of Rembau — and the bureaucratic dilly-dallying around it, plus the disruption of World War 1 and the Great Depression, was so glacial and soporific that the idea was just about lost.

The murmurs began again in 1922, with a spike in correspondence in 1926 between the General Officer Commanding Troops Malaya, and the Federal Government.

Governor and High Commissioner Sir Cecil Clementi brought the notion back to life in 1931, suggesting to General Officer Commanding (GOC) Malaya Command Maj-Gen Leopold Oldfield, to begin exploring the idea of a Malay Regiment.

Persistently in his ear were Sultan Iskandar (Sultan Perak), Tuanku Muhammad Shah (Yang Dipertuan Besar Negeri Sembilan), Raja Di-Hilir Sir Chulan ibni Sultan Abdullah, and Dato' Abdullah Dahan (Undang Luak Rembau). They were strongly in favour of the idea.

After many discussions, in September 1932 Oldfield got the official nod: 'Proceed with the necessary preparations with a view to forming (experimentally) a Malay regiment.'

An urgent bill was passed by the Federal Council. A budget of $75,000 Straits Dollars was allocated for this experiment with the aim of putting 25 recruits to the test for six months under Captain Bruce: The 1 Experimental Company.

A tiny three-and-a-half line notice appeared in the Social & Personal column of the *Straits Times* on 11 April 1933 announcing his appointment as 'Officer Commanding the Experimental Company, Malay Regiment.'

Bruce was thrown into the School of Oriental Studies in London to pick up the Malay language, probably under the tutelage of Margaret B Lewis, who went on to write *Teach Yourself Malay*. He then steamed out to Asia, making his way to Port Dickson, a sleepy coastal town in Negri Sembilan, 100km south of Kuala Lumpur. An old rubber plantation — at the 5th Mile, Coast Road lined by casuarina trees — had been allocated as their base. It had previously been used as a Volunteers camp, called The Haig Lines, with four simple sleeping huts and a gravel parade ground. There was not much else in terms of infrastructure, making do with what existed on the plantation.

Culturally, by 'Malay' it meant they followed the Muslim faith, habitually spoke *Bahasa Melayu*, and were born in the Malay states.

Physical selection criteria required them to be between 18 and 23 years old, with a minimum height of 5'3" (same as for the Indian Army, compared to 5'6" for the British Army). A normal chest measurement of at least 31" was required, with an expansion of 2".

'Good material, and keen,' Bruce said. They needed to be, because he had high hopes: 'They were going to be drilled and trained until their hands, feet, and hearts burned. They were not just going to be smart soldiers. They were going to be the smartest damn soldiers anybody in the peninsula had ever seen, heard or dreamt of. The same would go for their turnout, kits, and behaviour on leave. If anybody let the side down … it was letting down their own Malay people, their own Sultan, and family down.' A high calibre of recruit was found and selected — many from good families and others from modest rural *kampongs*, looking for a good opportunity. Nearly half of them were English speaking.

In front of him were the 25 recruits – including Raja Lope (the adopted son of the former Sultan of Perak), Mohamed Taib bin Jais and Ibrahim Sidek – selected from over 1000 applicants, many of whom had seen the opportunity advertised in their vernacular newspapers, such as *Warta Malaya.* Initially only suitable candidates from the four Federated Malay States were enlisted — the states who were actually funding this exercise — but the windows were opened wider to allow in others from the non-Federated States, too, to make it a truly Malayan outfit. Johor, however, was still excluded because they already had their own Johor Military Forces in place. (It was Johor way back in 1885 who'd first mooted then implemented the raising of an all-Malay force of native regulars shaped and equipped along British lines.)

While recruitment quotas were set by state and district, Perak was acknowledged as supplying the best quality recruits overall. None was more prominent and distinctive than 25-year-old Raja Lope (assigned regiment #25). He'd attended Magdalen College School in Oxford and later became an undergraduate at Balliol College, Oxford. More than that, he'd made a name for himself at athletics there, and was a keen soccer and cricket player, too. Given his position and background, although he was a private, he must've held a lot of 'soft power' with the locals, and proved a valuable bridge between them and the European officers with his easy-going nature.

The regiment was initially to be run on *askari* lines, meaning British officers overseeing local other ranks (a concept used widely in Africa since the 1890s). Bruce had three British staff officers, who'd arrived in Malaya ahead of him, his adjutant being Adj Capt Kenneth Exham (from the Duke of Wellington's Regiment), CSM E Field (from the Royal Lancasters), RSM Albert McCarthy (from the Volunteer Forces in Malaya), plus No 1 Clerk, Malaya-born Chinese civilian Lim Paul. Bruce considered the latter to be 'an efficient, loyal, quiet man, with a great sense of humour'.

Each side sized each other up. Bruce knew the way to galvanise the Malays was to make this 'their' regiment: something more local than just a British transplant.

He had a valuable sidekick in Ken Exham, who had been a lieutenant with the Duke of Wellington's Regiment in Singapore for a few years. His father was a Lt-Col, with attachments to the forces in Shanghai, previously, so, Asian exposure. Exham had married Joan — whose grandfather was Sir Robert Hamilton, awarded the KCB for his efforts in colonial administration in India, Australia, etc — in Singapore, and honeymooned in Java.

When the Duke of Wellington's were re-posted to the UK in 1931, Exham was seconded for service under the Colonial Office, and made Adjutant of the Straits Settlement Volunteer Force (Malacca Vol. Corps). So he had good local experience, and the 30-year-old took on the local rank of captain.

Field also had some local exposure, having been based at Butterworth with the Penang and Province Wellesley Volunteer Corp (receiving a long-service medal there) before signing on with the York and Lancaster Regiment.

With Exham, Bruce set about designing the regiment's uniform and insignia. Malay elements included two *harimau berjuang* (fighting tigers), a local *kris* dagger, an Oriental crown, and the colours green (being the colour of Islam) and yellow (being the colour for Malay royalty). British interests were represented by the colour red.

A motto *'Ta-at dan Setia'* meaning 'loyal and true' was agreed on, written in Jawi script, the Arabic-based system — going from right to left — for writing the Malay language. More colloquially, it might also be translated as 'last man standing'.

Next, how to localise the uniform? A green velvet *songkok* cap, white stiff-collared *baju* top and trousers, with a green-and-red silk *sarong* was deemed suitably smart for their walking-out dress. A local Chinese tailor, Kwong Yuen, was assigned to cut all the first uniforms and mufti for the recruits. 'The Sultans kindly gave permission for us to wear the short sarong and the *kampong* girls got an "eyeful" when they saw the new Malay soldiers on leave,' laughed Bruce, who was no stranger to flirtatious — and even downright licentious — behaviour himself. It was said the recruits developed something of a swagger. Their physical prowess had improved, with weight, height, and chest measurements all increasing measurably.

**

Indeed, when Bruce had been posted to Poona, India, for six years, he'd met a young British lass Mary (also known as Polly to her friends) at a regimental ball in Nainital one evening. He dazzled her, and others, with his flashy dance moves and things progressed rapidly from there. As a conversation with her son, John, later ran:

'Mama, how long did you know him?'

'For a week.'

'Where was I conceived?'

'On a lakeside under a tamarind tree. It was very romantic. You mustn't blame me, darling. He swept me off my feet.'

As those lakesides are rather steep, it must have been a particularly precarious coupling.

But then Bruce was not one to stand still, and had moved on to other postings around the empire, including New Zealand and even Sudan. But wherever he went, sport was his thing. Cricket, yachting, tennis. He'd always be in the centre of photos, indicating he was probably the team captain. And, of course, parties were a mainstay. The Sudan Club and the Grand Hotel in particular featured large in his social life where no doubt he'd swept other ladies off their feet. He ignored Mary's letters to support their son, many entreaties going completely unanswered. Perhaps not so much an officer and a gentleman, as much as an officer and a swordsman?

**

Friday prayers were always observed as a key part of the regiment's identity, and a mosque built on camp, along with an *imam* (religious teacher), Haji Ibrahim Yusof, appointed. He noted the keen interest taken by Bruce in his men's religious life.

But moulding this new regiment was hard work, and the other thing he instinctively felt he knew about the Malay character was the central importance of food. 'Good soldiers need good food,' he reflected. 'I think more recruits were obtained by the words *"Makan ada baik"* than by *"Ta'at setia"*,' he once said, meaning 'We have good food'.

Food might've been a more important driver than money. The Indian Army system paid locals about one third of what their British counterparts received (meaning a Malay private would be paid less than £3 per day in today's money). Barrack visits by parents and families always played up the food, to show their sons were being well fed and looked after, and that the British Army was a benevolent European organisation not to be afraid of. It was a very strong showing of *semangat kekeluargan*, demonstrating a key Malay concept of the spirit of a familial bond, a kinship and sense of togetherness. Also the cohesion of this diverse group — brought together from many

Malay states, as well as different British regiments — was easy to see. The Malays value this harmony, called *perpaduan*, of everyone mucking in together.

And then the serious square-bashing commenced, under RSM McCarthy who – despite his lean physique – had a parade voice 'like a saw cutting galvanised iron, and spoke fluent Malay with a cockney accent' according to Bruce. He introduced them to weapons training and tactical work. Commands were in English, instruction in Malay. The local men warmed to his sense of humour and sense of fairness, as he drilled them in their Gurkha-style hats and khakis. Apart from being fair-minded, with a keen sense of humour, Bruce considered him 'smart as paint'. McCarthy's extensive experience and leadership with the Royal Lincolnshire Regiment were instrumental in instilling British military standards, while simultaneously respecting and integrating Malay cultural values. As such they established a unique military ethos and 'flavour' for the regiment. His passionate following of Chelsea FC meant he probably bonded well the local guys over football, too, the code having been introduced to the locals in the late-19th century by the colonials. They took to it avidly.

Five months was the allocated training window, but within four months Bruce was ready to show his experimental Malay recruits to the world. In August 1933 the High Commissioner Shenton Thomas was invited, along with two staff officers, plus some English civilian planters from the state (possibly including Hugh Pilkington who planted rubber nearby in Negri Sembilan and received Christmas cards from Bruce), plus some others whom had been veterans of the First World War.

What astonished the onlookers that day – apart from the crisp white uniforms against the dark skin of the men – was when they performed a complex drill which included fixing and unfixing bayonets while on the march. This complicated exercise was seldom performed by units of the British Army, yet the Malays executed it flawlessly. 'Both in their appearance and the precision of their movements [they displayed] the very acme of military smartness,' trumpeted the *Straits Times* report. 'The good results attained so far have been attributable to the willing response made by the men as much as to good handling, organisation, and thorough training. It is sufficient to say that the first batch of recruits has done credit to their race and the Federated Malay States.' There was much hopefulness that this would lead to the Malay Regiment replacing the Burma Rifles. That was the perceived benchmark.

The High Commissioner was most impressed, immediately recommending an extension of the program, the expansion of the unit up to 150 recruits — a full-sized company — and promotion of Bruce to major.

And so the program grew and developed through 1934. For the first time the company gave a drill demonstration in the capital's *padang*, in front of Lord Robert Baden-Powell, the youngest major-general in the British Army since the Boer War (and founder of the Boy Scouts movement). Maj Bruce and Sgt Ismail bin Tahar barked their commands. One of their unique ones was their *terima gaji* drill. In *terima gaji* they held the right wrist with their left hand, as per Malay custom in receiving something, transferred the 'money' across to the left, bringing both hands down smartly, snapping to attention, saluting, turning, and marching away. All in perfect sync. 'They showed a precision and uniformity which spoke of meticulous training, strict training, and fine esprit de corps,' wrote the *Straits Times* of their fine display.

Adnan Saidi, now a 20-year-old with a boyish face and prominent ears, from the coffee estate area of Kajang, was voted best recruit. From the *kampong* of Sungai Ramal — a close-knit Malay farming community with a strong emphasis on Islamic practices — Adnan was the oldest of five siblings, and attended the local village school. He then briefly became a trainee teacher before this military opportunity popped up.

The Masjid Haji Mat Saman mosque, which is prominent in the village, sits on the sandy banks of the Langat River, thanks to its settling in the 18th century by Islamic scholars, and traders from Sumatra and Java (in modern-day Indonesia). Given that he was a normal *kampong* boy with good religious values and academically inclined, he was just the calibre of recruit the army was looking for. And these qualities were to stand him in excellent stead.

**

Success piled upon success, and one can imagine a fair amount of congratulatory back-slapping in the corridors of power. The regiment provided a 50-man guard of honour at the Federal Council's budget meeting, where the council was told: 'The Malay Regiment has progressed far beyond original expectations.' They would now aim for a battalion-sized unit.

Recruitment drives often took the form of officers and other ranks visiting various districts for a couple of days, often playing soccer friendlies against a local team, staging training exercises, and putting on a drill exhibition. This way many recruits were enlisted on the spot.

Five hundred acres adjacent their current camp was purchased for £75,000 from the Kong Sang Rubber Company. Further budget was allocated for construction of permanent barracks, officers' quarters, guard room, medical station and surgeon's quarters, and a new rifle range.

On 1 January 1935, it was thumbs up: the 1 Experimental Company officially became the Malay Regiment. Bruce was given a Lieutenant-Colonel rank as their commanding officer. The recruits were enlisted on a three-year engagement. Initially arms and equipment came on loan from the Imperial Government: they would now be bought for the regiment by the Federal Government.

There was an air of permanency about this now. The first experimental phase was over. The first test had been passed.

# Growing in Confidence

The regiment split into two rifle companies, a machine-gun platoon — wielding Vickers weapons — and a headquarters section, which comprised signallers, drummers, etc. It was beginning to look and feel more like a battalion. Another religious official Lebai Muhammad Ali bin Abdul Manan was appointed to help Haji Yusof. He found Bruce to be 'fully co-operative'in religious matters.

The passenger liner SS *Chitral* arrived from the UK in Singapore on 2 February 1935, and three officers appointed to the Malay Regiment disembarked: Maj JT Bretherton, Capt JRG Andre, and Lieut AFJ England.

John 'Jack' Bretherton of the Royal Warwickshires already had a Military Cross to his name, although how and when he won this is unclear. James 'Toby' Andre, 39, was formerly of the Lincolnshire Regiment, with a sharp nose, bushy eyebrows and a ready smile beneath his moustache. Both Bretherton and Andre were married, and their spouses (and 4-year-old son James in the case of Andre) joined them in Malaya. Lt England had come via the Penang and Port Wellesley Volunteers and Prince of Wales Volunteers. All were keen cricketers, Bretherton especially so, having played for the Army team before.

All would soon be in action with the regiment's cricket team, who played against the likes of Negri Sembilan Club, King George V School, Singapore Cricket Club, Malacca X1, and other regiments such as the Gordons. Maj Bruce led from the front snaring 7 wickets for just 21 runs in one match, top scoring the batting and smashing the winning runs off the last ball in unbeaten heroic fashion. Not to be outdone Lt England snared 8 for 40 in another match, and Bretherton 4 for 17.

A feature of the team is that it was not just the British officers involved. Raja Lope was a keen and good player, Pte Omar Baki a keen bowler, Privates Che Amin and Ismail, too. In just a few years, half the regular team would comprise Malays, with Othman Rani capturing 8 for 21 in one match, and Cpl Zaniel top-scoring with a half-century. So they didn't just play — they often starred in the team. Which would have been good informal team-building, with British commanders alongside local privates, plus the social side-effects of it.

Yet it was serious stuff, too. Raja Lope was facing up to Negri Sembilan state fast bowler JE Edmonds one time, when the ball smashed into his forearm, fracturing it above the wrist.

Over the years, regimental boxing teams, soccer, and hockey teams, would all be raised in a similar vein. Raja Lope was also selected as a forward in the Negri Sembilan soccer X1 to play in the Malaya Cup. He was known for his excellent distributing as well as attacking prowess … although he wasn't so popular when penalised in his first match which led to a 1-0 loss. But it does say something about his feisty nature, and he represented the state for several years.

Four additional British officers and five additional British warrant officers and instructors, and a quarter master — who arrived from England by mail boat — were added in February 1935. Sgt W Rich from the Royal Berkshires was the machine gun instructor, Sgt H Wraight (Wiltshires) as quarter master, and Sgt FW Jones (Border) as signalling instructor, CSM R Brown (South Wales Borderers), and H Beech (North Staffordshire), were also in the mix.

Bruce was adamant that all should wear the insignia of the Malay Regiment rather than their British home units as a show of esprit de corps and common identity to bond better with the locals. (This, even though they were not directly commissioned by the Malay Regiment itself.)

Speaking the local lingo was a big help towards this. An education officer was installed. Che Abu Samah — who would be teaching Malay to the foreign officers — was seconded from the Education Dept of Negri Sembilan. He was given lieutenant rank, and became one of the regiment's most devoted and loyal members. Bruce, who'd done that short course in England prior, had apparently picked up the language fast, Exham spoke 'tolerable' Malay, while McCarthy and Field were well-versed in it. The Malays were also instructed in English, general subjects held in both languages.

Appointment of newer officers came with a series of domino moves, as Lt England was promoted to adjutant, Capt Crozier (Buffs Regiment) moved up to company commander, and Maj Bretherton up to 2 i/c to Bruce; timely, as his boss was about to head off on home leave.

**

But all was not going swimmingly for all British NCOs. In the early morning of 6 May 1935, a Tamil was on the beach near the Tanaka Hotel, 1st Mile Linggi-Port Dickson Road, one of many burgeoning seaside resorts at Port Dickson, when he saw a body floating just a metre or so offshore. He summoned the proprietress of the hotel, an elderly Japanese woman. She immediately called the police. The body was soon identified as CSM J Gallacher, who'd been in-country for 18 months. He had a small two-inch wound on the side of his throat.

Gallacher had been reported missing from duty the previous evening, so the police started digging around for more information, and an inquest into this death held.

A picture started to emerge. He had 'lost' his wife in the UK some time ago (they'd presumably separated) and he missed his four children. His superiors noted this change of mood and, while he performed OK on parade, his demeanour was otherwise not deemed satisfactory for an officer. His colleagues had also found him increasingly aloof recently.

It seems he'd installed a Japanese *karayuki-san* in the local Magnolia Hotel, and kept her for the past year, visiting her twice a week.

On May 5 at 2am Gallacher arrived at her hotel room in a rather disheveled tiny three-and-a-half: clothes wet and muddy, knees bruised. He claimed to have fallen from his rental car while intoxicated. After sleeping for a few hours, he awoke at 9:30am, had coffee, and remained at the hotel until 6:30 pm.

In the meantime, he was presumably spotted there. Because Port Dickson policeman NC Halsey went up to her room to interview him. He found Gallacher dressed only in a kimono, sitting on the edge of the bed, and the Japanese girl was also there. Halsey explained that he'd been reported missing, and was to report to OC Major Bruce when he was found.

Gallacher told him he'd be returning to camp as soon as he was properly dressed, and left.

Meanwhile a search of Gallacher's room by Sgt-Maj FW Jones showed his bed had not been slept in, and there was a letter addressed 'To whom it may concern …' This immediately alarmed him, because he'd seen Gallacher 'a little morose' the evening before: he didn't join them when they went to watch a movie in Seremban. He told Maj Bretherton, who brought the matter up to Maj Bruce. Bruce searched the room and found a bundle of photographs tied up, with a note on an envelope stating 'To be sent home.'

Just then they bumped into Halsey, the policemen, who reported that Gallacher had been found in the Magnolia Hotel. Maj Bruce detailed Sgt Rich and Lt Field to go to the Magnolia Hotel and escort Gallacher back to camp. Bretherton arrived at the hotel to find the two arguing with the Japanese proprietress, who said Gallacher had gone off to get his hired car.

Bretherton and Halsey then decided to search at the Port Dickson Club, but Gallacher was not there either. So they put out an instruction to all units that the missing CSM was to be arrested under the Mental Disorders Act, due to concerns about his well-being (the only authority they had to detain him).

According to the Japanese girl, he'd left the hotel room at 6:30pm to get a car back to the camp, and that was the last time she'd seen him. The elderly Japanese hotelier had seen him looking 'unhappy' but he greeted her politely with *'Tabek.'*

And that was the last anyone had seen him before his body was found. Later, at low tide, a razor had also been found, confirmed by his Japanese girlfriend to be one that Gallacher had always carried with him, which caused the injury to his neck. But the official cause of death came down as 'suicide by drowning.'

His superior officer Bretherton voiced disappointment that Gallacher had not complained to him earlier about any challenges he was facing.

Four officers and four NCOS acted as pall-bearers at his funeral.

**

Posters blared: 'Army Life is a Happy Life! JOIN NOW!' Steps were in place to increase strength to 280 men. Massive infrastructure improvements had also been made, the Port Dickson camp now boasting two barracks blocks (to house around 350 men and NCOs), a parade ground, central reception area, guard house, garages for the mechanical transport, plus quarters for medical and other officers. Roadways to and around the camp were also improved, and everything just in time to be sparkling for the inspection by the new Governor and High Commissioner of the Straits Settlements and British Malaya, His Excellency Sir Shenton Thomas, fresh from a posting in Africa.

Port Dickson turned it on for Thomas, Lady Thomas, and their daughter Mary, with bunting and floral arches across streets of the downtown area. Chinese shopkeepers let off long strings of firecrackers.

Three companies of the regiment faced the saluting base, while off to one side a squad of new recruits — dressed in white, with Gurkha hats — watched expectantly. A march-past included military exercises and drills, once again their rhythm and accuracy impressing the official party, which included the British Resident for Negri Sembilan, John Hughes.

'To be a good soldier you have first to learn to obey orders, without thinking, and if you do that you will make good soldiers,' said Shenton Thomas. He stressed this point three times, phrased in slightly different ways, somewhat condescendingly.

In March 1936, Bruce went on furlough. He was given a send-off concert at regiment HQ.

In London he met a 'classic beauty' in the form of an actress 20 years his junior, Diana Hance. After a two week 'whirlwind courtship' they were married. Maj and Mrs Bruce sailed back to Penang in October. By now officer quarters had been built, and they settled into a large white double-storied house for the next two years, along with their newly-arrived baby girl.

Outside of his military job, Bruce busied himself with cricket (where he excelled as a wicket-keeper although he was an excellent bowler and batsman, too), and golf at the Port Dickson golf course. Diana was an ambitious photographer, and spent time capturing the nature, people, and wildlife of the area. One day she snapped locals who'd killed a large tiger on the edge of the golf course. (The fact that the regiment had a reputation for tremendous marksmanship was perhaps purely coincidental.)

More British officers were added, such as Bill Sanguinetti (Royal Hampshires), Bill Richardson (Royal Ulster Rifles), Ian Graham (Seaforth), Douglas Morris, and Melville Portal. Because officers' mess facilities were not yet in place, many had to find accommodation elsewhere in the town. Government holiday bungalows, especially in desirable places like Magnolia Bay — a pretty beach cove with emerald waters — proved popular.

But local officers were hard to come by, because they could not be created instantly.

The salary for second lieutenants was $75 per month. Inducements to study the reading, writing and speaking of English were put in place. The first Malay regular officers — coincidentally all Perak men — were grown from the ground up since Day 1 at Port Dickson, then given three months of additional training in Singapore with the Royal Inniskilling Fusiliers in 1936.

Second Lt Ibrahim bin Sidek, Ismail bin Tahar, Raja Lope Nor Rashid, and Ariffin bin Haji Sulaiman received their swords from Shenton Thomas — in extravagantly-plumed white helmet — at the Federal Building in Kuala Lumpur personally. It was the first time Malays would receive swords from the the British Government, and somewhat ironic given that the Perak War in 1875-6 saw Perak facing off against Britain. (And it might well have been Thomas who spoke to one of the Malay NCOs in very stilted pidgin English, according to Denaro, not knowing that Raja Lope was actually Oxford educated!)

A further honour was then bestowed on the proud regiment by their invitation to provide the guard of honour for the Federal Council's budget meeting opening ceremony, instead of the FMS Police.

Around this time Raja Lope married YM Raja Nor Zihan Binti al-Marhum Raja Harun Al-Rashid, and they started a family.

By now the total strength of the regiment — including British officers, Malay warrant officers, and other ranks — numbered 630. While the rank-and-file were all Malay, apart from Chinese admin guy Lim Paul, there were two Indians who worked as a storeman and a groundsman. Oh, and there was a barber serving the camp, who just happened to be Japanese. Quite a few more Japanese had appeared in the area of late, quietly working as photographers, pharmacists, barbers, and so on. No-one thought much about it.

**

The training and exposure came in handy when in March 1937 labour troubles in the mining sector boiled over. Around 6000 coal miners at the Malayan Colliery's Batu Arang site went on strike over pay and conditions. Initially the police were detailed to handle it, but when the 1000-strong crowd turned violent against the management who failed to satisfy their demands, the workers took to them with *parangs*, poles and crowbars. The mob was threatening a complete boil-over, and police shots killed and injured a few resisters. Reinforcements were needed.

250 members of the Malay Regiment were sent out to help maintain order: 'A' and 'B' Companies, and half of 'D' Company. Along with 225 Selangor Police, the crowd was soon subdued, with almost 40 agitators and pickets arrested. The crowd eventually dispersed, accepted a ten per cent pay rise, and went back to work.

Under Capt Andre and Capt Frank 'Freddie' Young MBE (formerly of the Royal West Kent Regiment) and their Malay officers, the regiment was later praised for the discipline and restraint it showed. It was to be an especially trying time for Andre, whose son James passed away at the tender age of six.

More glamorously, the regiment — the youngest in the Empire at that stage — also got invited to King George VI's coronation in England, sending a dapper delegation of two officers and two NCOs. Its international standing was growing. And always people talked about how smart they looked, and their impressive drill work. They now had Band and Drums, too.

A military tattoo was held at Farrer Park, Singapore. The Middlesex Regiment was called urgently to Hong Kong, so the Malay Regiment got an invitation at five days' notice to participate. RSM McCarthy and Capt Melville Portal together with Inche Ismail and Inche Raja Lope prepared the arms drill presentation of the 170-strong party. A large crowd was treated to a thrilling spectacle, including fixing and unfixing bayonets on the march, and the *'terima gaji'* move, came close to stealing the show from the Gordon Highlanders. The finale brought 2000 Singapore garrison troops together with the Malayan units, to shouts of 'God Save the King!', then the rousing national anthem closed out the night.

They spent a couple of weeks in the Straits Settlements Volunteer Forces camp at Telok Paku and Beting Kusa, near Selarang on the east coast, near Changi. Lt-Col Bruce and his officers hosted an 'at home' there, with Sir Shenton and Lady Thomas among the guests.

It seemed their star was rising: they beat the retreat in Singapore at the Esplanade, plus conducted a parade at a mosque.

They also had the honour of posting the guard at Government House there. And what a sight it was as 100 other ranks, with a single Malay officer as escort, drummed their way from the Singapore Volunteer Corp drill hall, along Stamford and Orchard Roads, to Government House, to relieve the Middlesex Regiment. TH The Sultan and Raja Perempuan of Perak, and HH The Sultan of Trengganu were at Government House with Shenton and Lady Thomas to witness this route march. Most importantly, this march-past was the first occasion on which the Malay Regiment was left entirely under the charge of its Malay officers. A milestone of trust.

Later, three-month stints in Singapore with the 2 Lancashire Loyals, now based on the west coast of Singapore, helped to fast-track other officer candidates later. Close co-operation in training was made with the Loyals, a relationship which would blossom over the next few years. As it built up to battalion size, officers such as Sgt Adnan were sent to Singapore for officer training programs. He returned as Second Lieutenant, then was put in charge of 7 Platoon, 'C' Company, 1 Battalion. He married an Islamic religious teacher, Sophia Pakih Muda (sometimes referred to as Sophia binte Pakir) from his hometown.

Families were also kept together with a married quarters constructed, and over 250 married soldiers moved in with their wives, and children for those who had them. A co-educational school was built on site, the Malay Regiment Children's School, with Che Janor appointed as headmaster. Officers (10 of whom now were Malay) and warrant officers of course got preferential treatment, with their digs on a raised hill affording them views of the Straits of Malacca.

All this helped to substantiate their recruitment posters which promised (in Jawi) 'More Men Required in 1938: Good Food; Good Pay; Recreation, etc.' These large posters were plastered across all states on public notice boards, outside mosques, etc. Additional jobs such as clerks, lorry drivers, and so on, were also advertised.

It was a compelling proposition: free food of the best quality (wives and children enjoyed free food too), free initial issue of clothing, free quarters, games and sports, and even cinema tickets at cheap regimental rates (eight shows for 50 cents per month, wives and children free admission).

A Standard 1V pass in Malay was required, but exceptions made where other fine qualities were shown. Same for height: if someone was shorter than 5'3" they had the opportunity to become a bandsman.

At this stage privates could earn base pay of $15 per month, corporals $18, with $26.50 for sergeants, and $38.50 for CSM. On top of these were monthly proficiency payments of 60 cents for education attainments, and 30 cents for good conduct. After two years' service this would be stepped up to $1.80 for privates. The engagement was for seven years with the Colours, and five years with the Reserve.

There was also the incentive of travel to various parts of Malaya, and an annual trip to Singapore.

For example, cadet officers did the regiment proud winning all four top places in a Bren-gun course held in Singapore in 1938. Also in Singapore, the regiment competed in the Combined Services athletic team championships, where teams from Blakang Mati Gunners, the Punjabs, Loyals, Inniskilling Fusiliers, Gordon Highlanders, etc, pitted themselves against each other at Gillman Barracks, in generally friendly rivalry.

As an example of the sports and games available, the 'C' Company soccer team under their captain Sgt Mohd. Sidek — with the very tall L/Cpl Khalit as goalkeeper — won the regiment's Ham Cup knock-out competition. Their CO Capt George Denaro was proud and happily posed with the team and their huge trophy for photos. Denaro was a swarthy moustachioed Maltese, formerly of the York and Lancaster Regiment, whose father was a colonel. A keen sportsman, Denaro was very involved and respected as a rugby referee in the district. (Other players like Lt Morris and Ian Graham were stand-out rugby players, who got selected to play for the Army team.)

Denaro had other reasons to be happy, too. His sister Edith got married to fellow Capt Michael Braithwaite of the Norfolk Regiment (attached to the Malay Regiment) in the Garrison Roman Catholic Church, Singapore. Denaro and Braithwaite were good mates on the cricket team, and the best man was Capt England, also a keen cricketer. This brought two military families together, with Braithwaite's late father being a major. They honeymooned at the Cameron Highlands.

The regiment's top brass — such as Bruce, Bretherton, and Andre and their wives, Portal, Denaro, and Freddy Young — from time-to-time were invited to lunch at Government House, where they mixed with other officers and visitors from the growing services in Singapore, and captains of industry. (By contrast, their future adversaries in the Japanese 18 Division were already hard at war in China, having been involved in the battles of Nanking (Nanjing) and Canton (Guangdong). A very different preparation, indeed, and perhaps telling in the final analysis.

That year the colonial government had a film crew working on a propaganda documentary *Five Faces of Malaya*. It was designed to show Malaya's diversity, and the races in harmony, the five faces being Malays, Chinese, Indians, Europeans, and the indigenous *Orang Asli* communities. The crew pitched up in Port Dickson, and filmed the regiment going through its drills, showing Bruce and McCarthy alongside their men, with Adnan Saidi barking orders and leading the men in smart precision drills.

Already there was a growing clamour for the Malay Regiment to replace the Punjabis, led by Raja Muda in the Perak State Council. Others — like long-time British Resident in Perak, Sir George Cator — were not so convinced they were ready.

**

You get the sense that even by the middle of 1938, there was nervousness that the Sino-Japanese War in China could lead to a wider regional conflict. The massive naval base in Singapore was now in operation, which simultaneously assured yet begged questions about regional security.

Col JJ Lewellin, Civil Lord of the Admiralty, was in Singapore for the grand opening of the naval base, and took the time to visit the New World theme park. 'If only we could have a grand new world ourselves, in which all races got on as well as do the different races and creeds that inhabit Malaya, what a grand new world it would be.'

Addressing the Association of British Malaya, Sir Richard Winstedt said: 'I certainly hope Malaya will not know war and I hope the lamentable struggle in the Far East may end in mutual respect and understanding,' he said. 'If peace is an idle dream, then at any rate I see Chinese and Malays, Indians and Eurasians donning their gas mask and doing their bit.'

They then proposed a toast: 'The prosperity of Malaya!'

**

After his five-year tour, it was time for Major Bruce to move on. He had over-achieved on the objectives, and was awarded the OBE. A colourful farewell was staged for him at which he looked around, feeling justifiably proud of the barracks buildings which had risen on the former plantation, the Officers Mess, and listening to their own regimental Band and Drums. But most importantly, 'We had 10 Malay officers, all chosen from the ranks.' RSM Albert McCarthy was awarded an MBE in assisting Bruce to bring the regiment so far, so fast.

The entire regiment plus fifty European guests turned out for his farewell. Bruce praised his men for their high reputation for smartness and efficiency. But he closed his speech by saying: 'Your most searching test still had to come under fire.'

Bruce himself was privately reflecting on that fact that 'the only test the Malays had not had was the test of war, and how they would behave was the unknown factor.'

He had created the youngest regular regiment in the Colonial Empire, with a high morale and only one single desertion in all those years. Sir Shenton Thomas, Governor of the Straits Settlements, honoured him by saying, 'I don't recall a better example of the right man in the right place. Bruce has left it a unit of which we are all proud and in whose future performance we have full confidence. If I were asked to suggest the secret of his success, I should say that it is the confidence in their officers which he has been able to inspire in successive batches of recruits. They have learned gradually to have confidence in themselves.'

That test of their confidence and competence was to come sooner than anyone that day probably imagined.

# On Your Marks

And so Lt-Col Bruce returned to his old unit, the Lincolnshires — relinquishing his local rank and reverting to his substantive rank of major — and was replaced by Maj Jack Bretherton as Officer Commanding, Malay Regiment. (He would assume the rather ostentatious name of John Talbot Bretherton-Hawkshead-Talbot when he succeeded to the estate after the death of a relative in 1942.)

Training was stepped up: everything from long route marches to specialised use of various mortars and anti-tank rifles. Given the tropical conditions, the route marches were back-breaking, especially with 80 pounds of equipment per man, plus rifle and food. Notably, 3 Platoon 1 Battalion once marched 104 miles in about as many hours, yet not a single man fell out.

Exham's tour of duty ended in early 1939, moving on to a staff position at Aldershot Command (he would go on to earn the DSO and later attain rank of major-general, his brother Robert becoming a brigadier).

A Bren Carrier platoon was also formed, under Capt Reginald Carter — a 23-year-old assistant at the Kedah Rubber Company in Bandar Bahru, Selangor — and sent off to train with the 2 Loyals.

Combined Services manoeuvres were also staged in Singapore, 700 men (including 15 officers — nine of whom were Malay) setting up their 100 or so canvas tents in the Chinese High School off Bukit Timah Road. They immediately got down to business with camp training, machine gun drills, etc, in the school grounds. Maj Andre conferred with GOC Malaya Maj-Gen Dobbie, an astute man who knew that the Japanese would attack Malaya and that they would use the overland route down the peninsula. If only others heeded his intelligence — he was soon replaced by Maj-Gen Lionel Bond who was of the opinion that, with limited resources, it would be better to bunker down and defend southern Johor and Singapore rather than try and do an ineffectual job across the whole peninsula.

The 1/2 Punjab Regiment had also come from Taiping, and the 2 East Surrey Regiment had shipped in from Shanghai (and would later be assigned to Malaya and the 11 Indian Division). With six battalions on the island at the same time, the place was abuzz. Some were alarmed that something was amiss, and the local media did their best to placate the citizens that this was just a training exercise, planned months ahead. *Nothing to see here*!

This was against an increasing backdrop of Hitlerism in Europe, and war now seemed imminent. In fact, a few months later Britain was at war with Germany. Even if it seemed a long way off across the oceans.

One effect this had is that the Regimental Colour could not be consecrated nor presented, owing to the war. Further it was the only colour of the regiment, because it had no King's Colour because the Federated Malay States were not a colony (technically, a protectorate).

Voices in the press called for an expansion of the local program, not just an importation of more Indian units. As a result, an additional company 'W Company' was formed in early 1940, and given a war station in Singapore, an *atap* (thatched roof) camp adjacent the oil tanks at Normanton.

Soon, the 2 Loyals and the Malay Regiment were bonded together into the 1 Malaya Infantry Brigade, and taking over W Sector in Pasir Panjang, which included strategic assets such as the Labrador area with its massive guns and the Jurong Boom on the west coast of Singapore.

At Normanton Camp, west across the Ayer Rajah Road from the Alexandra Hospital, the newly arrived regiment was settling in. Lt Adnan Saidi's family, which by now included two sons, moved down with him, and they lived off-camp in the nearby Pasir Panjang area. Maj Ismail Babu's family also moved into the nearby Buona Vista area.

Other matters of the heart were played out more publicly. Lt Douglas Morris — whose father was a major — was due to fly back to Britain to marry Joan (who he'd met in Malaya). The day before he was due to travel, war was declared in Europe, immediately scuppering all travel plans. Plan B was for her to join him out in Malaya. Permission for her to travel was denied due to war-time restrictions. His CO Lt-Col Bretherton (a cricket team mate of Morris's), and even Sir Shenton Thomas, had escalated this 'romantic battle', and next thing, Joan was in the news because this predicament had reached Question Time in the House of Commons! War Office permission was finally granted for her to travel — she flew by flying boat to Singapore — and they were married in St Mark's Church in Seremban. The bride was given away by Bretherton, while Lt Ian Graham was best man, supported by Capt George Denaro and Lt RNC Gudgeon (both keen cricketers). The wedding even made sensational local headlines: 'War Ban Heroine Weds Quietly at Seremban.' They then slipped off to the Cameron Highlands for their honeymoon.

**

The calls for expansion — spearheaded by the Commander-in-Chief Far East, and the GOC Malaya — finally bore fruit, increasing the Malay Regiment to two-battalion size. Sir Shenton Thomas's announcement of the decision to expand the regiment was warmly welcomed across the Malay community and media, with lengthy supportive editorials in the *Utusan Melayu* and *Warta Malaya*. These spoke of the fighting ability and loyalty of the Malays to the British Crown. 'We have the greatest faith in every Malay soldier in the defence of his country and of the King if, and when, the test comes,' wrote the *Utusan Melayu* editor, welcoming the wider share in defence — even internationally given the growing spectre of war — the Malays would have.

Bretherton returned to Port Dickson as Regimental Commander, while Andre (now 42 and promoted to Lt-Col, with Denaro as his 2 i/c) assumed command of 1 Battalion, and Freddie Young commanded 2 Battalion, with now-Capt Douglas Morris as his 2 i/c.

The regiment's policy of promoting from within the ranks — based on ability, strength of character, and leadership powers — were evident, often leading to a commission. By this stage, five of the original 25 recruits were officers, one was the regimental sergeant-major, one a company sergeant-major, two were company quartermaster-sergeants, and one a sergeant.

In anticipation of a Japanese invasion, the British began to reposition their forces. The Malay Regiment, especially the 1 Battalion, was mobilised as part of the 11 Indian Infantry Division to defend key positions in northern Malaya. They had been involved in some recce work, leveraging knowledge of local planters like Mohamed Yusuf Bangs (actually British, but he perhaps took on a Malay name out of desire to integrate, given he'd been planting in the Pergau area for 15 years). He was also a member of the 4th (Pahang) Battalion of the Federated Malay States Volunteer Force.

The work they were doing was to see if a Japanese advance down through Siam could be forestalled, before they even reached Malaya. The answer was probably 'yes' but Operation Matador was never triggered for reasons of bureaucratic bungling and not wanting to incite a diplomatic row with Siam by crossing the border from Malaya.

Meanwhile, a poignant scene was playing out at Singapore's main Tanjong Pagar Railway Station. Lt Adnan Saidi was there to farewell his family, who were being sent back upcountry for their safety. His wife, Sophia, eight and a half months pregnant with their third child, was taking their sons Mokhtar and Zainudin back to their hometown of Kajang. Mokhtar, the eldest son, was only four at that time. 'I did not realise that it would be the last time I would see my father. My brother and I kissed our father's hand before he left us. He did not say much. My father just spoke to my mother and just said, "Look after the children". I could see that my mother was very sad. She did not say anything.' They were not aware at that moment that it would be the last civilian northbound train out of Singapore for quite some time.

Exactly one week after formation of 2 Battalion — and the subsequent disruptive reorganisation and reshuffling of men and materiel required — Pearl Harbour was attacked and they found themselves suddenly at war with Japan.

Many of the ranks were civilians just days before and had hardly learned how to polish their boots let alone shoot a rifle. And still, locals poured into Port Dickson, some from the beleaguered north-eastern states of Kelantan and Trengganu. (The IJA 56 Regiment had landed in the former, technically hours before Pearl Harbour was attacked, and quickly established a beachhead.)

'A' Company was sent to Kuala Krai in Kelantan, to act as a last line of defence, if that was ever needed, to protect the railhead. 'D' Company was closer to home, at Sungai Bakap in Penang. Also notionally out of harm's way well behind the front lines.

But everyone was in harm's way for years to come ...

# Blitzkrieg and Backsliding

The Malay Regiment's 1 Battalion was among the very first to see action in the Asia Pacific theatre. Located in Kedah near the Thai border, Jitra was a key defensive position in northernmost Malaya. The Malay Regiment, alongside other British Commonwealth forces, was tasked with holding back the advance of Lieutenant-General Takuro Matsui's Japanese 5 Division long enough to allow for reinforcements further south to be mobilised.

The defensive line at Jitra was prepared hastily but was meant to be a formidable obstacle for the advancing Japanese forces. (Most thought the Japs were going to be forestalled by the pre-emptive Operation Matador over the border in Siam.)

The Commonwealth troops — 11 Indian Infantry Division including Indian, British, and Australian units under Maj-Gen Edward Murray — were deployed in the area. However, the defensive preparations were incomplete, and many positions lacked proper fortifications, exposing the defenders to a highly mobile and aggressive enemy who started as they intended to continue: *fast.*

The Japanese advance was swift, employing blitzkrieg-style tactics. Their tanks rolled through roads that the defenders had not expected to be accessible. The Japanese had air superiority, and their artillery and tanks added to the confusion and chaos among the defenders. Within three days, they essentially controlled all Malayan airspace.

Propaganda and mis-information was in full swing with the *Morning Tribune* newspaper on 10 December 1941 blaring: 'JAPANESE TROOPS IN NORTH MALAYA BEING MOPPED UP.' The article said only small pockets of Japanese had landed in the first place, and were being duly sorted out.

In reality, the Japanese launched a series of intense assaults, using their tanks and infantry to punch through the British lines. The fighting at Jitra was fierce, but the poorly-prepared defenses and the overwhelming Japanese superiority in terms of tactics, equipment, and numbers led to the collapse of the Commonwealth forces. After two days of heavy fighting, the defenders were forced to retreat under chaotic circumstances. The Malay Regiment, along with other units, withdrew in an attempt to regroup further down the peninsula.

The fall of Jitra was a psychological first-round knockout, and marked the beginning of a series of defeats for the British and Commonwealth forces as the Japanese continued their relentless drive down the Malayan peninsula at an average speed of ten miles per day.

Following the fall of Jitra, Murray's leadership was scrutinised, and he received a highly-polished boot up the rear, and was replaced by Maj-Gen Lewis Heath. The Malay Regiment continued to be involved in key defensive actions as the Japanese forces pushed southward, driving the British and Commonwealth forces toward Singapore, in disorganised retreats.

Some would call it a 'withdrawal complex'.

A couple of days after the fall of Jitra, the British forces, including the Malay Regiment, attempted to regroup at Gurun. But once again, the speed and coordination of the Japanese attack proved overwhelming. The Malay Regiment was engaged in fierce fighting but had to retreat when the line broke. The defence at Gurun, though brave, was unable to hold the Japanese advance, which often bypassed strongpoints.

Next came the battle of Kampar, during which New Year's Eve came and went with little or no reason to celebrate. One of the more successful defensive actions, the battle of Kampar saw the Allies mounting a more coordinated and determined defence. Maybe Heath had them back on track?

The town of Kampar, located in Perak, was strategically important due to its high ground and defensive positions. The British forces, well-positioned on the hills, were able to inflict heavy casualties on the advancing Japanese. The 1 Battalion Malay Regiment played a key role in defending these positions, demonstrating discipline and tenacity. This was home-ground for many of the initial Malay officers, who came from Perak. However, after four days of intense fighting, the defenders were once again forced to withdraw due to the risk of being outflanked. The successful delaying action at Kampar, however, bought valuable time for the retreating forces to regroup further south.

The Battle of Slim River kicked off on 6 January when the Japanese launched a daring night attack using tanks and infantry, catching the defenders off-guard, and all but decimating the 11 Indian Infantry Division. The speed and aggression of the Japanese forces led to a complete rout of the British and Indian forces stationed along the river. Despite their valiant efforts, the Malay Regiment, like other units, was forced to withdraw.

The significant loss at Slim River opened the way for the Japanese to advance rapidly toward Kuala Lumpur, then ultimately on to Singapore. The Malay Regiment, having fought in several key battles along the retreating line, continued to show resilience, even as the situation became increasingly desperate. They must've had awful mixed feelings as they felt their homeland slipping away from them, especially as they withdrew through Negri Sembilan, the state of their Port Dickson HQ base — their spiritual home.

**

The 2 Battalion companies were having a rough time of it too. Their initial blooding came at Sungai Bakap in Province Wellesley (now Penang), where they dug in to defend a section of the Kria River, coming under fire for the first time from land, and air bombardment from Japanese fighters screaming along at 100m altitudes with machine-guns blazing.

'D' Company thwarted a crossing of the river by Japanese troops using sampans. The Japanese used boats pilfered from Penang Island to leapfrog down the coast and stage surprise attacks from the rear of the 11 Indian Division units, sandwiching them.

'D' Company got caught up in the mass movement to the south. They boarded a train from Ipoh to Kuala Lumpur, coming under low-level bombing attacks several times. Lt Mohd Ali's platoon was assigned the unenviable task of guarding a convoy of 30 trucks loaded with high-explosives from KL to Kuala Kubu, a good 80km away. Then they had to stand guard at the Explosive and Ammunition Depot.

From Seremban the rest of the battalion headed back to Port Dickson. What a strange feeling, in such different circumstances, that must have been for them to return to their once-idyllic base camp. For RSM McCarthy — a veteran of so much drilling — now everything was real, not hypothetical, and they kept a sharp eye out for enemy landings in the area. Most of the wives were still there: Mercy Bretherton, Joan Morris, Kathleen Young, and Grace Andre, among them.

With the Slim River having been lost, the regimental depot — once the scene of those proud parades — was disbanded, and new recruits divided up and sent to each of the two battalions.

The foreign wives too were hastily bundled, with whatever luggage they had, into cars and driven down to Singapore, and passages arranged for them. They all got away on the Felix Roussel, one of the last evacuation vessels to leave, on the 6 February.

As 'D' Company skirted the coast down to Johor, they came under attack from Japanese planes. They cut across Gemas, Segamat, finally reaching Johor Bahru and crossing into Singapore on 14 January.

That same day, the Japanese controlled all land north of Johor. An unthinkable situation.

**

In the north-east Maj W Richardson had earlier given the orders for 'A' Company to withdraw from Kuala Krai — where they were blooded early, outnumbered by the Japanese pouring into Pengakalan Chepa in Kota Bahru — and they fought their way south in stages to Kuala Lipis in Pahang.

Lt Raja Lope and his company were guarding iron mines (Japanese ones, ironically) at Dungun at the outbreak of hostilities.

Together with 8 Indian Infantry Brigade, they had failed to protect the key airfields in Kota Bahru, Gong Kedah, etc, which meant the Japanese now could strike anywhere in Malaya and Singapore by air from these captured fields.

By 9 January they were given further orders to move south to Buloh Kasap in Johor, where Westforce — under Australian commander Maj-Gen Gordon Bennett of the 8 Australian Division — was trying to secure the major crossroads to ensure the safe withdrawal to the south. The situation was chaotic, fluid, and fast-moving.

A week later, orders came through to move further south to Segamat because Muar had fallen in a significant battle with the rampaging Imperial Guards Division. Mental and physical fatigue was taking its toll. Maj Richardson was one of a dozen evacuated behind lines to Singapore for treatment, and Lt DD Rennie now stepped up to take charge of 'A' Company.

**

If 'D' Company 2 Battalion thought they were now safe in the fortress of Singapore, reunited with their other 2 Battalion *kawan-kawan* (mates), they were sadly mistaken. They took on those new recruits, reformed, got some fresh kit, and after just one night in their Normanton Camp were back on the road north to Senggarang around Batu Pahat, Johor. Their mission: hold the town, and keep the strategic Ayer Hitam Road open.

'D' Company was deployed on the south bank of the Batu Pahat River, where the Japanese were threatening to make a crossing. The usual Japanese tactics of softening up by air, artillery, and mortars carried on, and the Malays exchanged mortar and machine-gun fire with them. Lance Corporal Napi bin Ahmad from Kuala Lumpur was killed in action around this time, leaving behind a widow Sapiah binti Kulop.

The courage of the local platoon commanders — Lt Ibrahim bin Alla Ditta, Yazid Amhad, and Mohd Ali — was on full display, as the endless barrages and strafing runs hit their lines. Their positive and encouraging attitude kept the morale of their men in check, a recurring theme in battle: the Malay NCOs were a vital conduit between the British commanders and the men, but most importantly were powerful and relatable role models. Perhaps none more so than RSM Ismail bin Baba who consistently demonstrated remarkable courage and steadfastness. His composure under fire gave his 1 Battalion men encouragement during critical moments.

But the situation was confusing because the Japanese were south of many of the retreating Allied units. *What the hell?*

Lt Ibrahim disguised himself as a civilian and struck out on several occasions to make sense of enemy positions and dispositions. He even bravely penetrated Japanese positions. On one such recce he spotted two British guns which had been hit and abandoned, but were not destroyed. He took his platoon out to recover them. For this he earned the Military Cross.

'A' Company was being run ragged with the west coast campaign intensifying in mid-January, devolving into 'every man for themselves' escape scenarios for many of the British, Australian, and Indian units, who were copping a thrashing around Bakri and Parit Sulong. Rennie's men were battered and bruised, unrecognisable as the young smartly-dressed men who'd impressed so many with their drill work. Southwards through Yong Peng they went, via Ayer Hitam — possibly waving at their fellow Malay colleagues with an *'Assalamualaikum'* or an *'Apa khabar?'* greeting as they passed through their lines — back to the bosom of Singapore and the Normanton Camp.

By 23 January, the Ayer Hitam-Batu Pahat crossroad was untenable. Batu Pahat was crawling with Japanese troops, and violent skirmishes erupted. Sgt Idris bin Achin and his 1 Battalion company were set upon. 'My post was surrounded and I was overwhelmed and taken prisoner,' he said. Lucky to be taken alive, he was kicked and slapped, and later forced with other Malay and British captives to help the Japanese push guns down towards Johor, and carry out other menial fatigue tasks.

'D' Company fell back to the Senggarang airfield. But then came the bad news that the Japanese had leapfrogged further south, outflanking their units by land and sea. Communications were cut. Orders came from the British commanders to destroy guns, trucks, and anything else useful to the Japanese, and make a run for Benut before they were all encircled. It was a dire situation, and no doubt many a prayer was offered.

Reaching Ponggor near Rengit the situation became clear: this was the end of the road. Two thousand Allies, mostly of the Royal Norfolks, 2 Cambridgeshires, 6/15 Indian Brigade, and the Malay Regiment, were cornered. Lt Mohd Ali teamed up with Maj Laman (Brigade Major of the Indians) and another Indian sergeant to try and reach a divisional HQ which they knew to be at Pontian Kechil. The problem was that it was nearly 50km south, and there was an infestation of Japanese between them and that HQ.

The plan was hatched to disguise themselves as fishermen and sail down in a sampan. They reached Pontian Kechil at night, but about 750 metres from the beach they were spotted. Voices rang out. But not Japanese voices. These were English voices. And they were giving orders: 'Fire!' Mohd Ali's men were taken to be Japanese! Desperate shouting back saved the day, managing to convince them that this rag- tag bunch of Asians in a boat were, well, British. *Phew!*

They explained the situation: two thousand men of the 15 Indian Brigade and 53 Brigade are pincered and cornered and face certain destruction at the hands of the Japanese unless we can ship them out.

Twenty-four hours later the Royal Navy deployed river gunboats, including HMS *Dragonfly, Grasshopper,* and *Scorpion* to Ponggor, and snatched the men from the swamps, ferrying the grateful passengers 100km to Singapore. This was done over several nights. It was Malaya's Dunkirk moment. The Malay company arrived at Clifford Pier, Singapore, at 2am on 29 January. 'All units were completely flogged,' reflected Capt Taylor, further pointing out their weapons were so caked with mud they were rendered totally useless.

Just a couple of days later, the Causeway between Singapore and the Malaya mainland was blown.

The homeland of the Malay Regiment was now lost. But all was not lost, because their finest moment was yet to come.

# Suddenly Shrinking Singapore

The Malay Regiment, now fully mobilised on Singapore island with a strength of 1400, was an integral part of the 1 Malaya Infantry Brigade under Brig George Williams, and tasked with defending critical positions on the western side of the island: W Sector. Here the beach was strewn with anti-boat defences and barbed wire, while the roads boasted anti-tank and anti-personnel measures such as timber scaffolds. Concrete pillboxes stood stark and sinister every 500m or so. While Allied troops were deployed across the west, central, and eastern parts of the islands facing across the water to Malaya, Percival's money was on the initial attack coming from the east, in the area of Pulau Ubin. Lt-Gen Arthur Percival was GOC Malaya, responsible for the defence of 'Fortress Singapore'.

But the western sector was richer in key installations — such as ordinance depots, barracks, oil supplies — and good potential landing sites.

1 Battalion was set the task of beach defences west of Singapore city, extending up to the Jurong River. To their immediate south were their good friends the 2 Loyals, on Blakang Mati (now Sentosa Island), protecting the southern maritime approaches and supporting the Malay Regiment's right flank.

Officers from the British Army General List were assigned to units that needed them. One such was Capt Howard Bretherton — no relation to the regiment's commanding officer — who had been working in education in Malaya, and was attached to the Malay's 1 Battalion. He'd previously been a Territorial Army officer in the Federated Malay States Volunteer Force. His wife Molly and 16-month-old daughter Merryn managed to snag a berth from Singapore to Australia, so he was relieved they at least had got away.

Capt (Acting Major) Griffith Richards from the South Staffordshire Regiment was also attached to the Malay Regiment, and their lives would soon be indelibly intertwined.

2 Battalion was still considered under-strength, with about one-third being fresh barely-trained recruits, and positioned along the Jurong River as mobile reserves, with forward elements west of the river, with the task of delaying

and disrupting any Japanese advancements coming from the northwest. Adjacent to their north was the 44 Indian Infantry Brigade.

We now know that Japan's 25 Army feinted an attack to the east, causing Percival to knee-jerk in that direction, then launched their initial attack landing the Imperial Guards on the northwest, at mangrove-ridden Sarimbun Beach.

22 Brigade Australian units there were quickly overwhelmed by the size and speed of the attack (which should not have been a surprise after observing the Japanese for the previous two months). But they were largely inexperienced — 'recruited Friday, sailed Monday' AIF Commander Maj-Gen Gordon Bennett called them — so the Japanese 18 Division — under the aggressive and highly-motivated Lt-Gen Renya Mutaguchi — broke through quickly.

The 44 Indian Infantry Brigade melted in the heat, proving ill-trained, ill-disciplined, and all too ready to cut and run. So 2 Battalion quickly found themselves face-to-face with the blood-thirsty Imperial Guards, and were outmatched. They were attacked as first light heralded the dawn of February 10. Intense combat led to the enemy penetrating their position, and massive casualties.

'After being in action for nearly three hours my company was ordered to withdraw to our Battalion Headquarters at Sungai Pandan,' recalled Lt Mohd Ali. They found themselves on the east side of the Jurong river. The preparations and protections of W Sector — such as pillboxes, mines and wire rolls — were abandoned. Brig Williams, ruling out possibility of a Malay counter-offensive due to insufficient reserves, ordered the Pandan Bridge to be blown.

2 Battalion was tired and stretched thin from the West Coast Road up along the Pandan River.

**

Meanwhile, a rather dispiriting spectacle played out on the coast: the balance of Britain's 18 Division was steaming down the Straits in a four-ship convoy. In sight of the Malay Regiment in W Sector, the last ship the RMS *Empress of Asia* came under fire from dive-bombers, was set ablaze, and burned for four days before it sunk. The sight of dozens drowning before his eyes would leave a harrowing impression on Pte Ujang Mormin, a 21-year-old from Kampung Kundur, Rembau.

**

The ominous rumble and roar of tanks clanking down Jurong Road sent shivers down everyones' spines. Behind and alongside them, Japanese infantry were hastening towards Ulu Pandan Road and Reformatory Roads.

Fighting was close-quarters and chaotic, the lines being jumbled and unclear. Australians and Malays and Indians and Japanese literally stumbled into each other.

Somewhere in this seething mess, the regiment's medical officer Capt William 'Bill' Young — who'd just married Kiwi lass Peggie the year before, and been with the FMSVF before being attached to the Malay Regiment — had been out helping and retrieving some wounded. He'd just made it back to the lines, when there was some incoming mortar fire. He jumped into a trench, landing on top of Doug Morris, as mortar shells burst around them. Two nearby were killed. Young was severely injured in the right arm, and rushed off to the Alexandra Military Hospital, where he was treated then re-assigned to help the overwhelmed staff there. The hospital was bursting at the seams with around 900 patients sardined into wards designed for less than half that number.

Also hospitalised there was Lt Jack 'Hugo' Hughes, a tough kind of guy. South African by birth, he'd moved to New Zealand to farm before moving to Malaya in 1930 as a 23-year-old. As an assistant planter he was doing his bit when he signed up for the FMS Volunteers. His wife, Frances, evacuated with her mother, grandmother, and their baby boy, Gerald, who was less than two. They made it safely to Fremantle. A great relief to him, but just two days later, he was fatefully attached to the Malay Regiment. In the cascade of mortar fire that rained down on these hills, he took a blast to the leg, and was rushed to 'The Alex' where his leg was amputated.

**

Those Malay infantry garrison troops at posts K5 and K6 on the west coast were relentlessly shelled, strafed and bombed as part of a general bombardment of Pasir Laba, which was home to a coastal defence battery. 'B' Company 2 Battalion had been the garrison troops for them since the start of the Singapore campaign. Now Capt RA Monks ordered his 130 men to get the hell out of the vicinity of the guns, all escaping miraculously unscathed. The guns were soon put out of action.

'B' Company headed south on the Jurong Road, joining forces with 'A' Company — who with 'C' Company 2 Battalion had been skirmishing with Japanese units around Jurong — and covered each other. Post-midnight on 10 February — under constant barrage from mortars and big guns — they reached the relative sanctity of their camp near the Brickworks at Normanton for some *makan* and finally some brief rest.

By lunchtime on 11 February the perimeter had shrunk even more. On the coast, Pasir Panjang was in friendly hands, and demarcated the frontline. Then it ran up and across to the crossroad where Reformatory Road met Ulu

Pandan Road, then across the Australian section to Bukit Timah Road, where the Japanese 5 Division — and its many marauding tanks — was held up.

Just as on the Malayan mainland, orders were given to withdraw south. 1 Battalion Malay Regiment was quickly repositioned to Pasir Panjang Ridge, a strategic spine of hilly land which controlled access to the installation-rich Alexandra area. It also offered a commanding view of Singapore city — and easy access to it — just to the south.

**

After sundown on 11 February, 1 Battalion 'C' Company was patrolling the village area near the Pasir Panjang crossroads, when 2/Lt Abbas' seven-man patrol stumbled across the enemy. Under the frantic shouted commands of Abbas, the company's heavy machine-gunners spat fire at the enemy patrol, holding them off ... for now. Lucky they'd evacuated their beloved families from here before all this happened.

To their east heavy shelling was softening up the 2/18 Australians trying to hold Reformatory Road near Holland Village.

The next day, the regrouped and resupplied Japanese put in an even more concerted attack. Wave after wave poured towards the position of the 1 Malaya Infantry Brigade. The Aussies withdrew to Holland Village, with 2 Battalion covering Ayer Rajah Road and the Gap Ridge. To their western flank were 'A' and 'C' Companies of 1 Battalion.

Singapore's survival strategy was a day-to-day proposition. But on the battlefield, it was more second-to-second, as bullets and shells and shrapnel whistled and blasted and boomed all around.

The Japanese were taking no prisoners, literally. Lt Muhammad Isa Mahmud of HQ Company 1 Battalion was captured along with other HQ Company officers. They were given the ultimatum of 'Join the IJA or else...'. Some acquiesced — in the heat of that desperate moment, why not? — but Mahmud didn't. He was executed for his stubborn resistance.

That night on Pasir Panjang Ridge the Malays could barely catch a wink of sleep with mortars thudding in and around their positions all night.

# The Battle of Pasir Panjang

The fact that the next day was Friday 13 would not have added any cheer to the situation, especially for the superstitious. How could things get worse? Well, they were about to find out. Percival had shrunk the perimeter again: now 'Singapore' was pretty much just Pasir Panjang Village to Macritchie Reservoir, across to Adam and Farrer Roads, and Kallang Aerodrome on the east coast.

Artillery, mortars, and fighter-bombers rained hell on the 1 Malaya positions, heavily huddled and concentrated and exposed as they were. A Japanese reconnaissance balloon floated above the battlefield, and called in the fire with devastating accuracy. At least five officers of the Malay Regiment were killed here, and many more other ranks men.

As proof of how fast the front line was moving, 1 Battalion's HQ — set up in Gap House — was obliterated, wounding Maj George Wort and an additional three officers and several other ranks. The Gap was important to hold, because it was a defile through which the enemy could easily pass through to the Alexandra Cantonment.

'A' Company 1 Battalion was already fighting for its life, when disaster struck. Its Company Commander Capt Horsburgh was killed in action, along with platoon commanders Lt Russell and Lt Haggit. Not only was their leadership gone, but so were their communication lines ... now just a mass of severed, burned and tangled wires, after the communications dugout was blasted out of existence.

Mortars were raining so fast and furious, the rate of fire could be mistaken for machine-gun fire, according to Lt Alistair Mackenzie, a 26-year-old insurance man who'd moved out from Northampton to Malaya four years earlier to work with Commercial Union Assurance. Now here he was, leading a carrier platoon.

Toby Andre was always popular for his calm and cheerful nature. On the battlefield, he proved his leadership mettle, springing into action. No comms? No problem! He — and his 2 i/c Denaro — ran and scuttled between their position and the forward troops, sending tactical messages but also revving them up in these loneliest and lowest of moments. 'C'mon, lads!' His easy smile came less easily in those moments, his bushy eyebrows knitted in stern determination. His leadership and gallantry here earned him a DSO. 'It was a tribute to the men, they way they fought,' he'd later say, humbly.

Lt Taib bin Jais watched helplessly as the 2 Battalion HQ — set up in what was nicknamed Durian Camp adjacent Pandan Hill, trig point 110 — was shelled and burned to the ground, causing countless casualties.

**

Survivors of this battle talk of the crescendo of noise. It kept getting louder and louder as the 'guns of February' from both sides traded blows, as though heralding the start of the imminent heavyweight title bout. Communications chaos reigned supreme, between units, between divisions, between HQs and frontlines. No one had any idea what was actually going on or who was really where.

At 1pm Mutaguchi issued the order: 'Attack Hill 270. Advance along Buona Vista Road. Attack Pasir Panjang Heights. Then advance to Keppel Harbour.' This set up one of the most pivotal battles in the battle for Singapore.

The Allies on this western flank were initially charged with defending the slopes and approaches to the Pasir Panjang Ridge. Around 2pm, under the tropical heat made worse by the smoke-blackened sky — literally a 'sinister twilight' in Churchillian terms — 2 Battalion was moving its chess pieces around this savage chessboard: 'A' Company came up, 'B' and 'D' fell back. At this inopportune time, the Japanese — lunging for the finish line of this war — threw everything at the Allies: Air. Tanks. Artillery. Mortars. And their troops came pouring in behind, sweeping across in an arc from Bukit Timah to Pasir Panjang village and Buona Vista village. Right into the Malay Regiment's sector in other words.

It was never expected that the full thrust of 18 Division would be thrown against the Malay Regiment directly. 13,000 seasoned Japanese versus 1400 untested Malays.

2 Battalion took it head-on for around two of the longest, most punishing hours that Lt Mohd Ali and CSM Abbas Abdul Manan could ever imagine. The grass around them burned and crackled and smoked, making it difficult to identify friend from foe. Until 2 Battalion could take it no more, and their commander Freddie Young ordered them to fall back to a pre-determined position near Point 270 below Reformatory Road.

A dozen British stragglers fell in with the Malays, as did Capt Poole and about 60 of his anti-aircraft gunners, who'd been displaced from their battery. They now took up arms as infantrymen.

Hill 270 aka Buona Vista Hill aka Pasir Panjang Ridge (now known as Kent Ridge), was the highest point of this coastal ridge at a 280m altitude. Strategic gold. That's why Mutaguchi wanted it so badly.

The 2 Battalion dug in to a hollow, and took effective toll against the Japanese 56 and 114 Regiments formed up with their tanks coming in behind them. While the Malays lobbed grenades — maybe all that cricket ball throwing was useful practice? — and fired their rifles and machine-guns with devastating effect — barrels becoming too hot to handle — from behind them mortar shells lobbed, and artillery hit their mark too. They staved the Japanese off for a while, buying valuable time for stragglers to make it back on the right side of their lines. 'D' Company Commander Taylor, and 'B' Company Commander Capt Monks made it to the hill, and Abdul Manan spotted Lt Yazid Ahmad, Lt Ibrahim Sidek, and Lt Masefield through the smoke.

The Japanese had set fire to the *belukar* (dense undergrowth) to expose and smoke out the defenders. Pte Yaacob bin Bidin, from Kampung Tebing Tinggi in Pahang, joined the dots of what might befall his 1 Battalion comrades, especially as he now noticed a mortar position being set up. With his Bren gun, using the smoke as cover, he leopard-crawled through the *belukar*, steadied the gun on its bi-pod and probably unleashed a full 28-round magazine at the mortar crew, putting it out of action. This brave action would earn him the regiment's only Military Medal of WW2.

'B' Company of 1 Battalion was also right in the thick of the action. One of the bravest here was Second Lieutenant Noor Din bin Ibrahim, who led his platoon with determination, contributing significantly to the defence of the ridge. His actions exemplified the resilience and valour that shaped how the Malay Regiment was beginning to perceive itself.

Under intense enemy fire, Corporal Abdul Rani bin May Hussein also exhibited exceptional bravery and leadership, earning a mention in dispatches.

The rampaging Japanese troops, emboldened by their artillery and armour, forced them to retreat. However, before the retreat could be made, the company was outflanked and encircled. Ammo supplies were depleted as they emptied their magazines at the Japanese, then fixed bayonets to take them on at close quarters. As Malay officers fell, Capt Yazid Ahmad — of the Federated Malay States Volunteer Force's 1 Perak Battalion — was seconded to this company to bolster them. He took command, leading them in a desperately heroic last stand before he was cut down where he stood, unyielding. Only a few of his men managed to break through the encircling troops and find a way out.

Those defenders captured were mostly lucky to escape with their lives, but given savage beatings and often put to work immediately doing menial labour for the Japanese.

'A' Company was slashed to ribbons, and the Japanese broke through, crossing Reformatory Rd.

Stragglers of the battered battalions withdrew after dark, heading towards the Alexandra Brickworks, and having to pass the roiling cauldrons of the Normanton Oil Depot — which had either been bombed by Japanese planes or lit by the British in a scorched-earth moment — on the way.

'C' Company was operating stubbornly at the coastal end of Reformatory Rd. Some Loyals, displaced gunners from the anti-aircraft battery, and Australians mucked in to bolster them, but they became isolated as the lines fractured and broke haphazardly towards midnight.

Luckily, 10 British Bren-gun carriers were dispatched to pull them out from their sticky situation. They sardined aboard these, choosing lulls in the mortar and artillery bombardment to dash through the lines. They were safe for now. But there would be no sleep that night.

1 Malaya Brigade and 44 Indian would withdraw to a line generally equal to Mount Echo — at the junction of Ayer Rajah and Depot Road — across to Buona Vista. Less than 8km from the downtown Singapore GPO. So close the Japanese could almost smell the victory now.

2 Battalion got the order to go into reserve, taking up a rear position near Mt Faber. They would see no further action, apart from patrolling Weir Wharf and the Mount Faber precinct for possible Japanese landings (the Japanese were mistakenly thought to have taken Blakang Mati island opposite it). Roadblocks were also set up and manned against a possible tank breakthrough. As a contingency they also recce'd positions for an entire brigade to withdraw to on the 'wrong' side of the Singapore River, which would act as a defensive moat in any last desperate stand.

Just after midnight of that fateful Friday 13, under the red glow of the oil fires, the rather battered but unquestionably resilient Malay 'C' Company 1 Battalion evacuated to a low hill west of the state-run Opium Packing Factory, some 50 metres inland off Pasir Panjang Road (these days, approximately at the site of the InterLocal Centre).

The Chinese referred to it as Opium Mountain; Malays knew it as Bukit Chandu.

## The Malay Regiment in Focus

A collection of photographs capturing the early years of the Malay Regiment, from its humble formation in 1933 to its elevation as the Royal Malay Regiment in 1947. These images offer a rare glimpse into the regiment's history, traditions, and service, and key personnel who shaped its narrative.

The 'father of the Malay Regiment' Major George Bruce, a decorated WW1 veteran. Note scar on cheek. Below: His wedding to Diana Hance in 1936, and their married quarters in Port Dickson, Malaya.

The regiment's insignia with motto '*Ta'at dan Setia*' in Jawi script, and their distinctive *songkok*-style cap.

One of the earliest photos of the 1 Experimental Company, probably in year of formation, 1933.

Below: Officers of the company in March 1934.

Front (L-R): RSM Osman Engkok, 2/Lt Ibrahim Sidek, 2/Lt Raja Lope, RSM McCarthy, Lt-Col Bruce MC, 2/Lt Ismail Tahar, 2/Lt Arrifin, 2/Lt Md Taib.

Back (L-R): CSM Ahmad Noordin, CQMS Manap, Mr Lim Paul, CQMS Musa, Sgt Ngah.

The regiment's Port Dickson base in 1936, on 500 acres purchased for £75,000 from the Kong Sang Rubber Company. Christmas card sent by Lt-Col Bruce.

2/Lt Raja Lope receives his sword from Shenton Thomas, 1936. Below: Former Malacca Volunteer Capt Ken Exham (front) with Sgt Adnan Saidi (top right).

In 1938 the regiment was featured in a propaganda film, *Five Faces*. Here, Drill Sergeant Adnan Saidi puts them through the paces of their signature drill work which so impressed audiences locally and internationally.

A Malay Regiment rifleman in standard tropical field uniform, circa 1940. The *kris* dagger was only worn for ceremonial purposes with the dress uniform.

Bayonet practice in Singapore 1941, where a lot of their advanced training was undertaken with British units. Below: around this time, coastal defences were constructed, like this pillbox on Pasir Panjang Rd on the west coast of Singapore.

Members of the Japanese 18th Division attacking Bukit Chandu -- a hill on the west coast of Singapore -- on 13 February, 1942. The brutal two-day battle came to define the skill and heroism of the Malay Regiment.

Homegrown hero Lt Adnan Saidi, killed in action at 27.

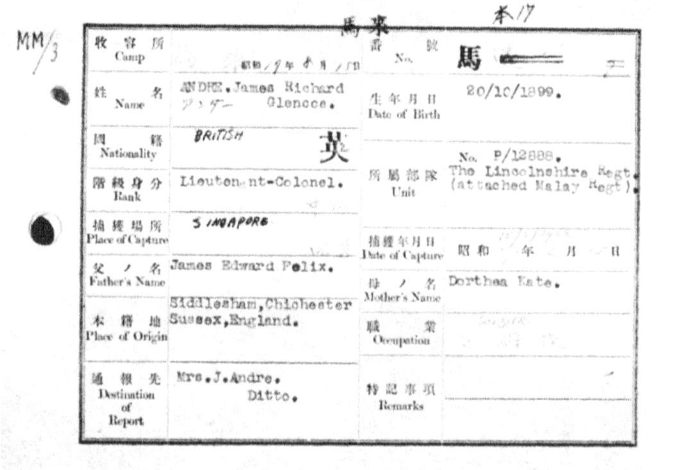

MM/3

本/17

| | | | |
|---|---|---|---|
| 收容所 Camp | | 番號 No. | 馬 |
| 姓名 Name | ANDRE, James Richard Glencoe.<br>アンダー | 生年月日 Date of Birth | 20/10/1899. |
| 國籍 Nationality | BRITISH 英 | | |
| 階級身分 Rank | Lieutenant-Colonel. | 所屬部隊 Unit | No. P/12888.<br>The Lincolnshire Regt.<br>(attached Malay Regt) |
| 捕獲場所 Place of Capture | SINGAPORE | 捕獲年月日 Date of Capture | 昭和 年 月 日 |
| 父ノ名 Father's Name | James Edward Felix. | 母ノ名 Mother's Name | Dorthea Kate. |
| 本籍地 Place of Origin | Siddlesham, Chichester Sussex, England. | 職業 Occupation | |
| 通報先 Destination of Report | Mrs. J. Andre.<br>Ditto. | 特記事項 Remarks | |

The POW card of Capt Toby Andre, who ended up on the Thai-Burma Railway. Below: Capt Howard Bretherton's Military Cross won 'for running away'.

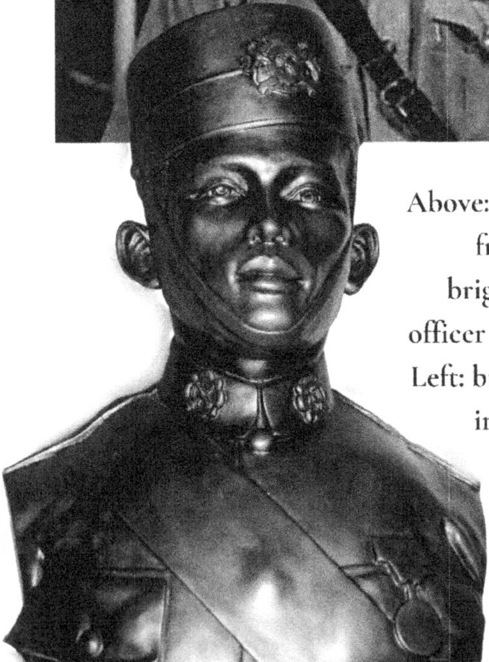

Above: Raja Lope who went from private in 1933 to brigadier, the first Malay officer to receive the salute. Left: bust of Lt Adnan Saidi in Reflections at Bukit Chandu, Singapore.

# Bukit Chandu Goes up in Smoke

The colonial government had a monopoly on the opium trade, relishing it as a major source of revenue. In 1935, the number of opium addicts in Singapore was estimated to be around 35,000, approximately 5 per cent of the Chinese population at that time. Immediately pre-war, due to various anti-opium measures and regulations, that number had reduced to 16,552 registered opium addicts.

The large red-tile-roofed factory on Bukit Chandu and the opium trade were part of the government's efforts to regulate and tax the consumption of opium. The factory processed raw opium into smokable opium. Licensed opium shops then sold the product to consumers, primarily targeting the local Chinese population, who were the primary users.

Hill 226, as Bukit Chandu was also referred, was a prominent bump in the landscape, covered by nutmeg trees and secondary jungle.

To the direct east about 3km away is the Alexandra Hospital, and the Alexandra Barracks and depots — seen among the distant treeline — would only be about a 15 to 20-minute march if unimpeded.

But there were lots of impediments. The Japanese had set the *belukar* undergrowth on fire on Chandu, which, combined with the smoke from the Normanton oil wells, meant that visibility was severely hampered and fighting was done in extremely close quarters.

On Opium Hill, 2/Lt Adnan Saidi – with his 42 men in Platoon 7 — was observing ragtag groups of Punjabi soldiers— around 20 or 30 of them — making their way slowly in ranks of fours up the slope.

Hmmm, why are Punjabis coming from that direction? Also, British units didn't typically march in fours, rather three abreast. As a former drill sergeant, he was immediately suspicious. You mean …? *Damn, Japanese disguised as Punjabis!*

'They were wearing disguises, some were wearing turbans to make themselves look like Indian soldiers,' said Abbas Abdul Manan, who was ready to feed magazines into Adnan's air-cooled Lewis machine-gun. He had several 47-round drums ready to go. As the men of Col Yoshio Nasu's 56 Infantry

Regiment reached the firing lines of the Malay Regiment, they felt the full brunt of 'C' Company's machine-guns, which could spit out around 150 rounds of .303 cartridges per minute.

Other rifles opened up too, finding their deadly mark. After all, four of the top marksmen in the 1938 military competition in Singapore were drawn from 'C' Company. Most of the charading Japanese died in the first fusillade. Others rolled or crawled down the steep hill to safety.

Adnan was laying down withering fire with his Lewis gun to devastating effect, and Abbas changed three magazines in that frantic exchange. Adnan copped a serious injury but kept shouting encouragement. '*Tuan*, if I should die today, I am quite willing as long as someone can look after my family,' he shouted to a fellow soldier. He was doing his best to live up to his own motto: '*Biar putih tulang jangan putih mata.*' This can be interpreted various ways, from the literal 'let the bones be white, don't let the eyes be white' — meaning remain steadfast in facing death rather than accepting the shame and regret of defeat which comes with not being steadfast. Alternatively, it can imply 'it is better to die fighting than to live crying in regret till the eyes become blind'. Or, more simply: death before dishonour.

Two hours later, the Japanese regrouped and launched an all-out assault, outnumbering the defenders on the hill by at least 10:1. The Malays were by now already short on ammunition and supplies, but fought with their all in keeping with their motto.

But it was more than that. It was their Malayness which came to the fore in those desperate days. The concept of *bersatu teguh* brought them together in these most challenging circumstances. United we stand. All for one and one for all: cooperative collectivism over individual glory.

Another cultural weapon they had was the *sorak*, a morale-boosting call-and-response war cry. '*Hu-ha! Hu-ha!*'

**

Pepys Road, named just a few years earlier for Johor General Advisor Walter Pepys, hosted a few bungalows dating back to 1910, built by the Colonial Government for senior employees, especially those in the shipping and opium industries. Reached via a series of hairpin bends, these double-storied 'black-and-white' colonial-style mansions were grand, spacious, and airy. The house at 31J was the residence of the Inspector General of Opium.

The equally stately house next door at 31K, was also once a grand residence, but had probably been requisitioned and seemed to house military and food supplies at this time.

'C' Company was at the pointy end of the defence line, when they set themselves in position at 31K Pepys Road. They would retreat no further,

Capt Harry Rix told his men. He would die with them right there, if needs be. He stationed Saidi's 42 men there.

Rix was made of stern stuff and a natural leader who'd been head of house at Sherborne School, a school prefect, in the 2nd XV rugby team, and a useful allrounder in the school's 1st XI cricket team. He'd gone on to study law at Cambridge and, because he'd gone out to be a solicitor and advocate in the Supreme Court in Malaya, found himself caught up in this action after he signed up.

The Japanese came hard at them, as they sheltered inside and around the houses at Pepys Road. 'There were many more of them, and they came from all sides,' recalls Abbas bin Abdul Manan. Grenades rained down like a tropical shower. But the Malay men stood fast.

Sometimes the Japanese got into the buildings. The Malay men repulsed them. Then they got in again. And were pushed out. Lt Stephen led a bayonet charge, but was shot down in front of his loyal batman Pte Wan Ngah.

Rix was cut down at the age of 41, just a day before hostilities ended.

'The Company Officer Capt HR Rix maintained close personal touch with his men, and showed disregard for danger which inspired his men to equal efforts,' wrote Dol Ramli, who interviewed survivors. 'He gave orders that the position should be defended to the last man and the last round, and the majority of the 'C' Company followed his example and died fighting.'

Command of 'C' Company 1 Battalion passed automatically to 27-year-old Lt Adnan Saidi.

**

'B' Company was doing its best to hold the road junction at Buona Vista Village, with platoon commanders like Lt Khalid bin Hashim proving their mettle. They were bouyed by three Aussie carriers. But these proved no match for the Japanese tanks, who put in a concerted shove around 2pm. Seeing one carrier damaged, the others withdrew. Sgt Hussein Mat Som of 'D' Company was sent with his Vickers machine-gun — no match for armour — to bolster 'B' Company, but was soon injured.

At 1600 hours the Japanese troops probing away at the left flank — ably supported by their tanks thundering along the coastal Pasir Panjang Road and Bueno Vista Road — eventually forced the defenders back off their line. Lt MacKenzie was at 'B' Company HQ when the company commander Capt (Acting Maj) Griffith Richards had gone forward up the road in his Bren carrier to investigate shooting coming from his forward platoons. A hundred yards up the road, a Japanese tank opened fire from 50 yards off, hitting the carrier twice and disabling it. MacKenzie — a very keen rugby fan — calmly leapt into action, scurrying along the hedges and walls that lined the road, to rescue the commander and crew. Some shouting from the roadside attracted his attention, just in time to see a Japanese crew bringing an infantry gun to bear, supported by a tank. He levelled his Tommy gun, emptying his magazine, killing or wounding the entire gun team of about 15, allowing the commander and his crew man to make it back to safety, uninjured. This effort won him the Military Cross.

Mind you, he was almost mown down by friendly fire as he made his way back to his own lines.

Not so lucky was Capt Reg Carter, 33, commanding the Bren-carrier platoon of 1 Battalion. His carrier was hit, probably by tank shells, and he was wounded, evacuated, but would die of his injuries nearly two weeks after the surrender.

From the Mount Echo-Bueno Vista line, they were now pushed back to the Alexandra Brickworks, and even along the canal to Bukit Chermin ('Mirror Hill') near the Labrador guns on the coast. Many 'B' Company stragglers made it back to Labrador. Amid the confusion, with the coastal guns deemed to be in enemy hands, the Allies' artillery attached to the Malay Regiment was dropping 25-pound shells on the area, shelling its own men. *Terima kasih!* Pte Hamid bin Man, Pilus bin Aris, and Yassin bin Haji Latif were killed by these shells, or skirmishing Japanese units.

Lieutenant Abbas bin Mohd Saiid played a pivotal role in the defence alongside Adnan and his soldiers. As an NCO, his leadership and courage were instrumental in resisting the enemy's advance, despite being heavily outnumbered and facing overwhelming odds. Tragically, Lt Abbas was killed in action during this fierce battle. As was Sgt Ali Hashim, who was mentioned in dispatches for his gallantry in this final action. Ditto, Pte Abdul Wahab, remembered for his bravery during this battle, fighting tenaciously to delay the Japanese advance. Alongside his comrades, he engaged in close combat, defending the ridge and Opium Hill until he was killed during the final stages of the battle.

Finally, after a desperate battle with the Malays standing and holding their ground, the Japanese broke through. All Malay officers, with the exception of Lt Adnan and 2/Lt Abbas bin Abdul Manan, had been killed in the vicious onslaught with close-quarter exchanges of grenades and automatic rifle fire.

All around bayonets from both sides flashed and clashed.

Pte Ujang — affectionately known as 'Tok' — had been in action alongside Adnan, when artillery shrapnel smashed into his rifle, blasting it to pieces, sending shards and splinters into his hands.

But the Malay Regiment was about to pay very heavily for its gallantry in frustrating the 18 Division advance, and delaying its movement southward to be the first to reach Singapore town. For that was the trophy that lay tantalisingly close to the blood-thirsty troops of Mutaguchi's men. They were in a heated competition with their counterparts in 5th and Imperial Guards Divisions. And it wasn't a friendly rivalry.

To claim Singapore for the Emperor was their single-minded focus.

**

Soon, the Japanese were savagely mopping up. Lt Ahmad Noordin of 'A' Company 1 Battalion was executed, once again for probably refusing to join the IJA, just hours before hostilities ended. Cpl Din was also captured and tied up, along with five others. The enemy then bayoneted them, and left them for dead. Badly injured, once dark, he crawled away, chancing upon Pte Wan Ngah, who helped him back to their lines. Other captured survivors were massacred, and Cpl Yaakob, in the chaos of the carnage, fell on top of a pile of dead soldiers, taking the chance to lie motionless and play dead. He unwittingly witnessed a gruesome spectacle.

This involved the hero Adnan, who the attackers would have recognised from his short stature and his manic machine-gunning. They beat him savagely then hung him upside down from a tree. Then came the call to bayonet: 'Yell! Stick! Turn! Retract!' as the Japanese bayonet drill went. Several thrusts, deliberately designed to give him as slow and painful a death as possible.

Two days later, upcountry, his wife Sophia gave birth to their daughter, but the baby died soon after.

The only officer who survived that brutal bout on Bukit Chandu was Lt Abbas bin Abdul Manan who, with four of his men, fought their way backwards.

'C' Company became separated from 'D' Company — near Labrador — and they had no idea where 'B' Company had got to. They just wanted, needed, to get the hell out of here. *Now!*

But suddenly they were faced by a seething wall of flames. It was a wide drain — up to 20-feet wide in parts — which ran down from Normanton Oil depot, and was full of burning oil. This flared up into an impassable flaming screen, 'a wall of fire' according to Abbas.

Cornered on three sides, there was nothing for them to do but attempt to jump it. Two fell short in their desperate attempt to ford it, pulled through with ghastly burns to most of their body. They bravely crawled away to hide, whilst the other three made good their escape. (Mackenzie later would come across these two, 'wounded and covered in oil'.)

In something like a scene from an action blockbuster, the bullets continued to sizzle through the fire at the fleeing group. Abbas and the other few survivors were able to scramble back to 1 Battalion HQ, and report the grim status of his lost company.

**

Around dusk — although frankly the entire day seemed like dusk due to billowing clouds of smoke — a careless moment by the perhaps complacent or prematurely celebrating Japs gave a chance for retribution. Overlooking the Alexandra Brickworks area, 'D' Company and their battalion HQ — plus an assorted mish-mash battalion fused from 5 Bedfordshires and Hertsfordshires, and the badly-mauled Leicestershires well-armed with 3-inch mortars — saw two companies of Japanese openly marching down Pasir Panjang Road. Definitely 18 Division, probably 56 or 114 or 124 Regiment.

MacKenzie's men couldn't believe their luck: 'Imagine our surprise and delight when the Nips appeared marching straight down the Pasir Panjang Road — in fours! We let them go to within 100-150 yards of 'D' Company HQ and then let them have it with machine-guns first and then with mortars.'

Pte Sulong Ahmed reckoned it was with a lot more than that: rifles, Lewis guns, rifle grenades, anti-tank rifles, and even revolvers.

'Almost every man of the first company was slaughtered and a lot further back too,' McKenzie reckoned. 'A Loyals officer up on the right counted 94 bodies on that part of the road that he could see alone.' A further 300 were wounded. It was like a fairground shooting gallery.

From the roadside ditches, any surviving Japs were firing back. Ahmed was seriously injured in the neck, and taken to hospital. With the Alex Hospital — immediately to their east — now in some kind of no-man's land, any wounded were evacuated to Singapore town itself in six of the Loyals' trucks arranged by Capt JG Jesson, in the absence of any ambulances.

And so ended the Battle of Bukit Chandu. The Malay Regiment had carved a legacy for itself. They had staunched the Japanese 18 Division's final west coast thrust towards Keppel Harbour with possibly the fiercest resistance the Japanese had faced in all of the Malayan Campaign. But too little, too late. A Pyrrhic victory of sorts, yet a proving ground for the men and the very idea of the Malay Regiment.

Battles are one thing, but the war itself was not yet over.

# An Eerie Silence

Sunday, 15 February, 1942. Thankfully, with the Japanese held in check for the time being, 1 Battalion had time to regroup and lick their wounds. The Malays were holding positions in front of the Singapore docks across the back of the Keppel Golf Links, to Bukit Chermin Road and Morse Road.

The occasional shell lobbed into their positions, but the focus seemed to be a little to the east, where the concentration of military infrastructure was: Gillman Barracks, the ordnance depot, and the Alexandra Hospital which — unknown to anyone elsewhere — was witnessing a series of horrific massacres at that time.

Capt William Young, the Malay's medical officer-turned-patient, worked tirelessly tending to the worst of the worst, even as the Japanese rounded up all other medical officers and patients and mercilessly 'disposed' of around 250 of them. Any survivors were made prisoners of war.

The Loyals, who had been so closely associated with the Malay Regiment for many years, were now fighting for survival in their very own Gillman Barracks.

Further back on his hilltop look-out, Mutaguchi was growing wildly impatient with his 18 Division's stalled progress. The 114 and 55 Regiments' line remained practically static on account of heavy firing from the direction of Keppel Barracks. Nasu's 56th was also pinned in the same position as yesterday, dug in near the railway line.

The British kept lobbing 6- and 9.2-inch shells overhead all day, into Japanese positions around the barracks and hospital.

'The British Infantry appear to be on the verge of surrendering,' thought Warrant Officer Arai Mitsuo, 114 Regiment, 'yet their artillery keeps pounding us. They should all be shot.' The 114 — desperate to reach its target of Hill 136, Gillman Barracks — crawled forward through the rubber trees on their stomachs, shells bursting around them. The Loyals HQ Coy, under Maj Francis Barnes, used these fusillades to retreat to Mount Washington around 16:30 while the enemy was pinned.

Capt PVS Holden (attached from the Sherwood Forresters to the Malay Regiment since 1938) was in an armoured car that came under heavy fire from a Japanese patrol and he died, just hours before the surrender. Lt TPD Jones (who had started with the Singapore Volunteer Armoured Car Company before being attached to the Malay Regiment) was mentioned in dispatches for his gallantry during this campaign, and survived … for now.

In the general fracas, Captain Howard Bretherton (attached to the Malay Regiment) was captured on what turned out to be the last afternoon of hostilities, the day Singapore fell. At first, they would jam him into the top floor of Gillman Barracks in overcrowded conditions, together with Major Griffith Richards (also attached to the Malay Regiment). Here they'd remain for four intolerably long, hot, sweaty days and nights.

The Loyals were digging into their new positions on the high ground above the barracks around Henderson Road and Mount Washington when some Japanese suddenly appeared. The Loyals' 'A' Coy fired on them, possibly the last shots fired in the battle for this scarred and smoking southern island. 'They'd already bombed the hell out of Singapore' said Pte Stan Sharpley of the 2 Loyals. 'There was nothing to stop them, nothing at all.' Nothing, that was, except an unconditional surrender.

'At 20:00 suddenly the Army HQ made a call to the Division HQ saying "The enemy surrendered", said the Japanese *Senshi Soshi* battle report. 'The 18th Div shall soon terminate your combat action, stay where you are tonight, and maintain the highest alert toward the enemy.' Arai sensed a sudden commotion from the back moving up to their front. 'It seemed to be coming from the coast, through the valley, rising up the hills, like a growing tidal wave to engulf us,' he said. 'The roar grew closer and louder. One could not see their silhouettes but the force of their voices was shattering: *Banzai! Banzai! Banzaaai*"

The guns fell suddenly and eerily silent. The Battle of Singapore was over.

**

Lt Gen Percival later heaped specific praise on the 'the very fine performance' and actions of the Malay Regiment: 'The Malay Regiment showed what esprit de corps and discipline can achieve. Garrisons of posts held their ground and many of them were wiped out to a man.'

The Malays have their own equivalent term for esprit de corps: *gotong-royong*. This concept emphasises mutual cooperation — communal effort and collective responsibility for the greater good. A shared struggle for survival.

# Extreme Loyalty

The survivors of the Malay Regiment reassembled on Monday, February 16, in the Keppel Golf Links area, together with their Commonwealth allies, whom they'd fought shoulder-to-shoulder alongside. Their brutal war was over, but this was not the finish line yet.

Upon announcement of the surrender, Lt-Col Andre had ordered his 1 Battalion Malay officers and men to disperse and disappear into the general population, thus pre-emptively saving them from what was about to unfold.

Tok Ujang was one of his men who took off his uniform and headed off to blend into the local populace as a villager. Luckily a Singaporean Malay family took him in. For the Fujiwara Kikan, an intelligence agency under Maj Fujiwara Iwaichi, was about to deliver a final tragic test — and twist — in the story.

The F-Kikan gathered intel, mainly by fostering alliances with local nationalist movements on the ground in Malaya and Burma pre-war. Indians in exile and Indian expatriates were typical targets.

During the Malayan campaign itself, the Japanese played a masterful hand of propaganda. Leaflets dropped on the Indian lines targeted them with messages along the lines that they were just the cannon fodder for white man's ambitions in Southeast Asia. It played to many Indians' feelings that they were second-class citizens even in their own country's army, and certainly their quality of life in Malaya was nothing to fight for. They were encouraged instead to defect, and they would be set free. They could hand in their arms and uniforms and in return get a pair of shorts, a bowl of rice, and — importantly — their freedom.

It transpired that 32-year-old Capt Mohan Singh Dib, of 1/14 Punjabs — who surrendered upon the fall of Jitra in the earliest days of the conflict — was one such turncoat who switched sides, and thus became the notional founder of the Indian Nationalist Army (INA). These men were pressed immediately albeit informally into action, still wearing British-issued uniforms plus a Japanese armband, fighting against their former comrades down the Malayan peninsula and on to Singapore.

So on Tuesday February 17, all 'native' POWS were directed to assemble in Farrer Park — scene of the thrilling Military Tattoo in 1937 — which had large open sports fields, and was ironically near the Little India precinct of Singapore. The Malay Regiment was to be included in this, and the eight surviving Malay officers and 600 men were separated from their British officers and marched off.

Most of the British officers went to the Changi Gaol precinct, the sick and wounded to Roberts Barracks.

Among their British officers, Commanding Officer Lt-Col Bretherton, Maj Andre, Maj Denaro, Capt Morris, and medic Lt Young, would end up on the Thai-Burma Death Railway. Popular cricketer from Middlesex, Capt Gudgeon (son of a vicar), would die on the railway in mid-1943. As did Armoured Car specialist Lt TPD Jones, who died of cholera in Songkurai. Meanwhile Acting Major George Wort, Capt Ken Exham, and RSM Albert McCarthy spent their POW days languishing in Singapore.

Let's also not forget those Malay Regiment soldiers who were taken captive during the battle of Malaya itself … they would nearly all end up as *romusha* (labourers) working in the most inhuman ghastly conditions — far worse than the other Allies endured, and with a far higher attrition rate — on the Thai-Burma Railway, and further afield.

Andre kept his focus on the future, dreaming that the Malay soldiers would once again parade on the square at Port Dickson, fortified by the knowledge that he and they had faced the supreme test of a soldier — which the battle of Pasir Panjang surely was — and passed it with flying colours.

Maj Fujiwara addressed the crowd of around 45,000 Indians, rolling out the whole 'Asian Co-prosperity Sphere' spiel, and how important an India free of British influence played a part in this framework. Then he handed over to Capt Mohan Singh, transferring command of the Indian POWs to him. After his emotive entreaties, a massive majority of Indians had by now defected to or claimed allegiance to the newly-formed INA, and sided with the Japanese. Those who didn't wilfully join were 'strictly punished' in Japanese parlance — marched off, blindfolded, and bayonetted into trenches.

The Indians had defected almost en masse, but would the Malays follow suit?

One did immediately: Maj Mustapha Hussein of 2 Battalion decided his future lay with the F-Kikan. But the rest, No. 'My husband Ibrahim Sidek and his friends were told by the Japanese to take off their Malay Regiment uniforms and accept release,' said Sharifar Khadijah binte Hamid. 'They refused. They didn't even want to remove their badges of rank.'

Hussein worked on the Malay captives, along with Singh, to get them to come over the victorious side. Still a resolute, No.

**

Meanwhile back at the Gillman Barracks on 19 February, noticing the guards were not paying attention, distracted by a car accident nearby, Richards and Bretherton took their chance, slipping down the stairs of the barracks building, and hid in an unused room. Once night fell, they crept out of the barracks area. With all Europeans rounded up and marched to Changi already, it was not easy for a couple of round-eyed officers to avoid detection by the Japanese patrols, as they lurked around, trying to scavenge food and plan an escape.

After three days on the run, they'd made it to the Pandan River estuary, where they spotted a *bedar*, a 12-foot traditional Malay sailing boat. They planned to sail that night, and set about supplying it with whatever essentials they could scrounge. Around midnight, they slipped quietly out of the estuary, running the risk of execution on sight if discovered.

By dawn, they came ashore on a small island, where — through the rather cagey and sceptical villagers — they ascertained they had reached the Dutch East Indies. The Merlimau Islands in the Riau Archipelago (in the vicinity of Batam and Bintan), about 50km south of Singapore. With their Malay boat, and scavenged bits of local clothing such as Malay hats and coats, they were able to slip past the watchful Japanese naval vessels who had wreaked havoc on the flotilla escaping Singapore in the previous week or so. A couple of local fisherman probably looked harmless enough.

Four days after leaving Singapore, they reached the more significant outpost of Karimoen, where Dutch authorities helped them connect with ships and escape parties — mainly fleeing expats from Singapore and Malaya — making good their journey to India.

Bretherton and Richards were two of only very few who managed to get off Singapore island once the Japanese had claimed victory. They won the Military Cross for their efforts (though sadly Richards would be killed in action in Chittagong, Bangladesh, in 1944).

**

By 28 February, the Japanese had run out of patience with the stubborn hold-out Malays (including those of State Volunteers). Such a show of loyalty to their former masters was beautiful in its innocence, brutal in its consequence.

At the galling urging of traitor Hussein, some 98 Malay Regiment personnel serving under the Imperial Forces were taken by the Japanese military police and machine-gunned at dawn at The Gap, Pasir Panjang, near to where they'd valiantly staunched the Japanese flow for a while.

'The Japanese executed them,' said Lt Sidek's wife, Sharifar Khadijah binte Hamid. His body was never found, and her only keepsake from her devoted husband was his army-issued tin mug. 'My children were still very young, too young to know their father,' Sharifar said. 'His body was never found. We cannot even remember him with a grave.'

Apart from Sidek were four hugely brave and loyal officers who accepted death before dishonour by not accepting release into the general population: Lt Mohd Ariffin bin Haji Sulaiman, Lt Abdul Wahid Jidin, Lt Abdullah Saad, and Lt Abbas Mohd Said. Gallant and patriotic to the last. While the rank-and-file foot soldiers were the backbone of the regiment, these NCOs were pivotal in maintaining the morale and disciplined fighting spirit of the regiment going in its toughest, torturous, most trying moments. In many ways they consistently set the tone of the narrative for the Malay Regiment.

**

A general announcement had been made in *The Good Citizen's Guide,* Syonan-to, advising all refugees who didn't belong in Singapore to scarper. 'Any refugee who is found in Syonan after March 14 shall have his property seized and shall be severely punished.' (Read: disposed of.)

There had been up to half a million such refugees pushed southward ahead of the Japanese invasion. Now they all headed back home under threat of death.

Lt Mohd Taib bin Jais, one of the 1 Battalion officers who had dispersed upon the surrender, returned home to Port Dickson in early March. But the Japanese soon caught up with him, looking to exact retribution. He was rounded up and thrown into the Teming Gaol at nearby Seremban with three other ex-Malay Regiment members. The latter rank-and-file soldiers were released shortly after, but the lieutenant was singled out for harsh treatment because he'd been an officer. He was beaten up and interrogated.

Then one day he — along with a trio of Chinese prisoners — was frogmarched out to the courtyard. Here they were forced to kneel on the ground, their hands tied behind their backs. Two Chinese were beheaded in front of him with a samurai sword by the Japanese prison officer. The interpreter desperately implored him to save the other two.

Unimaginably, the prison officer stopped in the middle of this gruesome task, allowing Mohd Taib and the other Chinese a stay of execution. They were marched, shaking, back to their cells. Then a few days later, Mohd Taib was released from gaol with a stern warning to behave himself … or else. Unfathomable behaviour. And what relief!

Tok Ujang was still in Singapore, a few months past the Japanese deadline. He hatched a risky plan to return home to Rembau. So one night he slipped into the main railway station at Tanjong Pagar, then snuck his way into the cargo compartment of a train headed for Malaya. He probably hopped off at

Seremban, leaving him about a 20km journey back home to his family in Rembau.

Raja Lope had also made it home to Perak, and during the 'Japan Time' Occupation was a crucial go-between for the Sultan and Lt-Col Peter Dobree of Force 136, who had parachuted into Upper Perak to organise the resistance and co-ordinate guerrilla activities. Raja Lope had quickly established contact with him, and for his involvement to have been discovered would've attracted summary execution from the Japanese.

**

During the defence of Singapore, the Malay Regiment lost 'at least' 600 men who were killed or died in Japanese hands. Nine British officers and seven Malay officers suffered the same fate.

And thus ends a very brave and inspiring saga for the local men of Malaya. They may have been vanquished, but stood proud as proof of concept for a rather bold experiment: The Malay Experiment.

Could Malays make an effective modern fighting force? A resounding *'Boleh lah!'* Yes, can do!

# From the Ashes

The end of the war brought the British running back to Malaya, and with them came the challenges of reconstruction, and, well, reconciliation.

An editorial from *The Times* in London gushed: 'British Rule Vindicated. Malaya's Great Welcome.' Their reporter in KL wrote: 'At every town and village on the road here from Singapore the people have built triumphal arches with welcoming inscriptions. The arches and the houses are decked with the flags of Great Britain, China, and the Malayan People's Anti-Japanese Army. Children still cheer wildly every passing vehicle, making the thumbs-up sign.' That part might be objective reporting, but then he veered off into more subjective editorialising: 'Perhaps under the Japanese, the common bond of suffering and hatred for the invader tended to unite the communities. Although physically separated by barbed wire, European and Asiatic have probably been closer during the past three years than before.'

But what was missed here was that — although Malayan people were often divided along ethnic lines — their abandonment by the British, and struggle at the hands of the Japanese, had forged a collective nationalism. *What doesn't break us, only makes us stronger.*

How could the people of Malaya trust the so called *tuans* — the British 'bosses' — anymore, after they had basically run a scorched-earth policy retreating from a country that wasn't even technically theirs.

In reality, the country was teetering on the brink of revolution, with a nationalistic push for independence headed by Ibrahim Yaacob. Yaacob had founded the *Kesatuan Melayu Muda* (Young Malay Union) in the 1930s, one of the earliest political organisations spoiling for independence in Malaya. One of his ideas was for Malaya to join with Indonesia, which declared independence from Holland in August 1945.

The Malayan Communist Party (MCP) counted itself as 5000-strong now. Some of their units even boasted European liaison officers. In some northern and central states Japanese commanders tried to broker deals with the MCP to make men and arms available to continue their battles. The lines were becoming very blurred.

But there was an irrepressible and irreversible anti-colonial groundswell.

Exactly one month after the Japanese surrender, on 5 September 1945, the Malay Regiment was revived. Lt-Col KS Hayes Palmer was temporarily in charge of reorganising the regiment. During their captivity, Maj Toby Andre and Maj George Wort (who would eventually reach brigadier rank) had separately taken the time, effort, and risk to compile the personal particulars and experiences of those who served in the Malay Regiment. This helped enormously to understand the performance and fate of the body of men individually and collectively. In 1946, these were further elaborated and refined as more information and clarity came to light, the casualty statistics more horrific than at first thought, resulting in the printing of an In Memoriam notice in the *Straits Times* to all those officers in 1 Battalion who fell.

Of course many of the local men had ended up toiling on the infamous Thai-Burma Death Railway. Within a few days of the Japanese surrender, a small party made a complete tour of all labour camps in Thailand, using a train they'd confiscated from the Japanese. One of this party was Capt AN Ross of the Malayan Civil Service, attached to the Malay Regiment. They brought medications and made arrangements for the labourers to be fed and looked after until they could be repatriated.

They found 26,000 Malays were still on the railway, while at least 100,000 were estimated to have perished already.

On 1 March, the anniversary of the regiment, the unit's *imam* conducted a ceremony at the mosque, involving three companies of war veterans. Special prayers were offered for all the officers and men who died fighting the Japanese, as well as those who died in their prisons and squalid work camps. A sombre moment of reflection.

Of the British officers, CQMS Oldfield had returned to Malaya post-war, and was enjoying life as a civilian in Kedah, a north-western rice-bowl state. The injured medical officer Doctor William Young became a radiologist in Singapore post-war, becoming a consulting radiologist in KL in the 1950s, then worked in the Dept of Radiology at KL University Hospital (which he'd do until the mid-70s).

Alistair Mackenzie also returned to the insurance game in Singapore, ultimately heading up Commercial Union for Southeast Asia, and would be awarded a Commander of the Order of the British Empire (CBE) in 1967 for his services to that industry.

Howard Bretherton (who'd escaped to India) returned to Malaya to continue his work in education. He had been involved with the Colonial Office in setting up an education system in Malaya throughout the 1930s. His post-war contributions further strengthened the educational framework in the region.

In a mark of the man, one of the first things he'd done in returning to Malaya post-war, was to try and seek out the the fisherman whose 12' boat he had 'borrowed' from Sungei Pandan to make his daring escape, so that he could repay him the cost. But he never spoke about his wartime experiences with his family, and died when he was just 58. 'It must have affected his health,' said his daughter Merryn. 'As for his Military Cross, he said he won it for running away.'

Maj Denaro returned to Malaya almost immediately post-war. He married wife Francesca in England 1946, and was back in Malaya as Lt-Col Commanding Officer of the regiment with a big job ahead.

Because in the aftermath of World War II, the Malay Regiment found itself at a crossroads.

GOC-i-C Malaya Command Lt-Gen A Galloway's main challenge was to maintain security for British economic interests and oversee the political process amidst post-war demobilisation. But while cautioning about potential security gaps, he saw the ongoing viability of the Malay Regiment as a matter of cost. Malaya would probably have to chip in to finance accommodation, organisation, and equipment to re-build the regiment. 'While Malaya had cause for pride in this regiment, it's future is still under consideration,' the WW2 veteran said.

The Japanese occupation had ravaged Malaya, and the regiment had been severely depleted in both numbers and morale. For the Malay Regiment, this period of rebuilding required a redefinition of their role. No longer were they a mere auxiliary force under British command; they were now being positioned as the primary military force in a country on the verge of gaining independence, a storyline being played out in several colonies across Asia simultaneously as the once-mighty British Empire crumbled. The resilient Malay Regiment would need to rise from the ashes of war, growing not just in size but in importance to the post-colonial narrative of Malaya.

One of the first moves was the 1 Battalion moving to Kuala Lumpur — under canvas in Mount Estate, Gombak — under the command of Maj GEC Ash. Assuming a peacetime role, they were assigned to guard the Governor's House, the Pudu Gaol, and other government installations. They were there to protect against a general lawlessness which had pervaded in the limbo since the end of the war. That it was Malays and not British or Indians in that guardian role was yet another milestone achievement for the regiment. The regiment also learned the importance of peacetime and normal urban discipline being as equally important as that of wartime. A new attitude.

British military advisors continued to work alongside Malay officers to build a more self- sufficient force. For the first time, Malays had reached the rank of captain: Ismail bin Taha, Raja Lope bin Rashid, and Taib bin Jais, were all part of that first intake of 25, and all saw active service in Singapore against the Japanese.

The ever-ready Lt Raja Lope had signed back on immediately and was involved in preparing the HQ depot for training. He was soon appointed ADC to Edward Gent. Gent explained that this way he'd have a direct and personal link to the affairs of the regiment.

In any case, 19 May 1946 was when their second chapter started, coinciding with the visit of Sir Edward Gent, the first Governor of the newly-created Malayan Union. (The Union became the Federation after just two years, and Gent was killed in an air accident when returning to London.) All personnel — except those previously discharged or medically unable — were notified to report to the regiment on or before 31 August or be struck off the strength of the Malay Regiment.

One of the ideas mooted was to open the regiment to other races, such as Chinese and Indians, to create an integrated multi-racial force, representative of the country itself. The Malay rulers, generally sultans, and the general population thought this idea stunk worse than a rotten durian.

Was this another cunning British ploy to erode and dilute *ketuanan Melayu* (Malay supremacy)?

By mid-1946, that idea was off the table. The Malay Regiment would remain just that: all Malay, within a Federation Army of one division, which would free up the British regiments for more strategic operations, while the focus of the home team would be more on domestic internal security. (The multi-racial idea would be raised again 20 years later by a Chinese politician Dr Tan, but be rejected.)

A contingent of 141 volunteers and regulars from Malayan units was selected to march in the grand Victory Parade in London. Toby Andre was selected to command them in the parade, accompanied by Maj Ford and around 30 other NCOs and other ranks, including Lt Ibrahim bin Alladitta, Sgt Noordin, and Cpl Wahab. Many were astonished that there were so many people in London: 5,000,000 spectators — about the total population of Malaya then — waved and cheered, 500 vehicles lumbered along, and 300 aircraft buzzed in the overcast skies overhead. The highlight for many of the Malays was catching a glimpse of Winston Churchill. The King himself had requested they wear the *songkok* instead of the Gurkha hat, and later asked questions about the uniforms and rubber. The round trip by boat meant they were off for around three months, and enjoyed accomodation in the Kensington Gardens area.

In a hint of the increased localisation that would gather apace over the years, four Malays received commissions at their new KL HQ: RSM Mohd. Nasir bin Yaacob, CSM Ali Hashim bin Hashim, CSM Mohd. Nor bin Abdul

Rahman, and Sgt Hussein bin Hassan were all commissioned by the outgoing GOC-in-C Malaya, Lt-Gen Sir Frank Messervy. He read out a statement in Malay talking about the dispiriting delays of getting the regiment back on its feet. 'But I can now fairly say that I am sure there is a great future for the Malay Regiment. My reason for saying this with confidence is because you, Malay soldiers, have the character and ability and keenness to prove yourselves as good as any in the world.' In signing off, he wished the regiment 'a glorious future'.

**

But there was a slightly inglorious side for some veterans. A Malay Regiment corporal was arrested in KL in a trap set by police who were trying to find the source of Bren gun and Sten gun ammo which had turned up in recent raids on gangsters' hideouts. A sting was set, with the undercover detective posing as a gangster, and the corporal handed over 1346 rounds of ammunition and three hand grenades for $2000. He was promptly arrested.

Down in Singapore, a similar case at the time involved Abdul Samad bin Jaffar, who had served with regiment and seen action in Dungun, Trengganu. He was caught in a house in Arab St with a fully-loaded .38 revolver, two Japanese automatic pistols, and 77 rounds of ammo. He claimed it was planted on him, but was sentenced to 18 months imprisonment.

Wahid bin Jumaat found himself in police custody one night, charged with armed robbery and causing grievous hurt, after he and three Chinese accomplices hailed a taxi, then forced him to drive them to Changi point at revolver-point. There the driver was bundled out of the car, robbed of $35 cash and some clothing, then Wahid shot him in the arm at point-blank range. The police chased the taxi, and arrested the foursome. Wahid was clearly suffering from what he said he'd endured as a POW on the Thai-Burma Railway and also in New Britain (Dutch New Guinea, now Western New Guinea, Indonesia). He was sentenced to 10 years imprisonment.

His story was only half-believed. But more came out when a Malay soldier from Malacca attached to the Royal Artillery was returned from Java with the help of British authorities. He reported that there were several members of the Malay Regiment still in Java. Some, he said, had already married Indonesian wives, and were generally getting on fine. Some arrived back with wives and even children in tow. Others were found in Rabaul, a major Japanese base in the South Pacific. Yet others were found in Hollandia, Dutch New Guinea.

It turned out that there was a 'lost legion' of Malayans, numbering around 230, who had been shipped to work as forced labourers in Java by the Japanese. Lt Mohd. Noor of the Malay Regiment was on the screening committee to vet and verify stories. Eventually eight members of the Malay

Regiment — including Pte Ismail bin Lest from Tampin, and Pte Bachee bin Mohd Shah from Alor Gajah — as well as a dozen FMS Volunteers were identified, the men given $5 pocket money and put on the night train back to KL.

So it was clear that the horror of war had clearly and understandably derailed some who served.

**

In January 1947 a familiar face re-appeared: it was Lt-Col Andre, recalled to Malaya as Commandant of the regiment. Who better than the man who'd stood tall and proud with his men in the darkest moments on Bukit Chandu?

He set about raising and rehabilitating 1 Battalion, then focussed on re-raising 2 Battalion.

Soon, an expansion of the regiment to six battalions by 1950 was endorsed.

Better still, the Malay Regiment was granted the title 'Royal' on 1 March 1947. The honour was bestowed in recognition of the regiment's outstanding bravery and service during World War II, particularly during the Battle of Singapore and its homegrown heroics on Bukit Chandu.

A significant ceremony was held at the Sungai Besi Camp, near Kuala Lumpur. As part of the ceremony, King George VI presented the regiment with its first Royal Colours (a ceremonial flag), symbolising the sovereign's trust in the regiment's loyalty and bravery. A grand military parade was held, featuring soldiers from the Malay Regiment marching in full ceremonial regalia and showcasing their discipline and skills, in front of an appreciative audience of high-ranking officers, colonial officials, and distinguished guests.

The proclamation conferring the title 'Royal' was read aloud, officially renaming the regiment as the Royal Malay Regiment *(Rejimen Askar Melayu Diraja)*. One can almost sense the chest measurements of all regiment members expanding with pride. *We did it!*

The humble Malay Regiment, started from 25 raw recruits, now had the royal stamp of approval. And what they had achieved in terms of establishing the identity, values, and brave behavioural blueprint of the regiment, would be the template of the further tests that lay directly ahead.

# Emergency Measures

It was a small headline in the bottom corner of the front page of the *Singapore Free Press* on 16 June 1948 that would send Malaya into a tailspin: 'Stop Press: 3 Planters Shot Dead Near Ipoh.'

Armed men ambushed the plantation and murdered three European estate managers in the Sungei Siput district, about 20km from Ipoh: Arthur Walker, John Allison, and Ian Christian, had just returned from their morning rounds for breakfast. The invaders used Sten guns and revolvers to effect the killings, searched and robbed the bungalows for any valuables, then set the rubber smoke-houses on fire.

Gurkha troops — probably 2nd Battalion of the 2nd King Edward VII's Own Gurkha Rifles (the Sirmoor Rifles) who'd just arrived in country three months earlier and were stationed in Ipoh — were dispatched to the scene, and all European women and children in the area were evacuated.

A few days later, police arrested a 'Chin Piang Nam', a 24-year-old Chinese labourer on the estate. The attackers were members of the Malayan Races Liberation Army (MRLA), the armed wing of the Malayan Communist Party (MCP), who sought to overthrow British rule and establish a communist state. These brutal killings were intended to send a political message against British colonial rule and their economic interests in Malaya.

What was the response? The British colonial government declared a state of emergency across Malaya on 18 June 1948, marking the start of the Malayan Emergency. This conflict became a protracted guerrilla war between Commonwealth forces and communist insurgents, involving significant counterinsurgency efforts.

The Malayan Emergency — *Darurat* in local lingo — marked the next critical phase in the regiment's history. This brutal, protracted conflict from 1948-60 saw the regiment face a new kind of warfare: not against foreign invaders but against domestic communist insurgents.

The story is evocatively told in *The War of the Running Dogs* by Noel Barber.

The MCP under Secretary-General Chin Peng sought to establish a communist state in Malaya by overthrowing British colonial rule. Ironically, the MCP had formed the Malayan Peoples' Anti-Japanese Army (MPAJA) during the period of occupation from 1942-5.

Was Chin Piang Nam actually Chin Peng? They were the same age, and the name sounds way too similar to be coincidental. (The specific individuals responsible for these murders were never identified, captured, or brought to trial.) Chin Peng was born Ong Boon Hua and became a guerrilla aged just 15, joining the MCP. Throughout WW2, Chin fought as an anti-colonialist guerrilla in the MPAJA against the Japanese occupation.

Under the premise of 'my enemy's enemy must be my friend' the British had supported them and even supplied them with weapons: WWII-era rifles, machine-guns, mortars, and explosives. The same weapons which were now being pointed back at them!

Chin Peng had aligned with Force 136, the British-funded covert resistance movement in Asia, famously connected with Lim Bo Seng and Freddie Spencer Chapman (who wrote *The Jungle is Neutral*).

Chin Peng was awarded an OBE at the end of the war. And, as the senior surviving member of the guerrillas, set about establishing the Malayan Races Liberation Army (MRLA).

New CO Lt-Col Bill Blair-Oliphant came from a very distinguished military family in Scotland (his father was also a Lt-Col, DSO, and his brother an Air Vice-Marshal).

Destabilising guerrilla warfare in the dense tropical jungles of Malaya tested the regiment's adaptability, endurance, and willpower. This tied in with a change of uniform to the new darker jungle greens. Veterans with wartime experience were often called back into service or joined up as trainers, making their skills invaluable during this period.

While 1 and 2 Battalions were rushed out into the field, square-bashing and intense training was going into turning more raw recruits into 3 Battalion. RSM Mohamed Isa bin Mahadi whipped them into shape, earning an MBE for his unstinting ongoing efforts.

Posters and handbills were printed and passed out to *penghulus* (village headman), for distribution to *ra'ayat* (citizens) in the various outstations they planned to visit.

But unbeknown for a while was the fact that a recruiting racket was going on. It was found that some youths who were selected as suitable soldier material, and issued with a free railways pass to reach the training depot before a certain cut-off date, had on-sold their eligibility to other aspirants who had been rejected. So a different person with lesser ability had turned up pretending to be the approved applicant on the list!

After that, Recruiting Officer Maj DA Banks — assisted in outreach efforts by medical officer Dr GA Lopez and Sgt Hussein bin Sulaiman — insisted an ID snapshot was taken at application, along with fingerprints, and checked when the potential recruit arrived. How many slipped through the net will never be known.

George Denaro was posted to the large garrison base in Sungei Patani, where recruits were being trained and put through their paces. His son Arthur was born in Sungei Patani in 1948, would also go onto a distinguished career in the British Army, reaching major-general rank.

**

Douglas Morris and wife Joan had come back to Malaya post-war. As a Lt-Col Senior Staff Officer, he would become 2 i/c to Toby Andre. He saw the Malay units taking well to jungle warfare over the next four years they were in-country.

This involved postings to Kuala Lipis in Pahang, with much of the work involving long-range patrols looking for bandits. Being in the central highlands, more rugged terrain and denser jungle conditions would be hard to come by, making their task all that more challenging.

The regiment adapted to this new form of combat, using their intimate knowledge of the terrain to their advantage. Alongside British and Commonwealth forces, the Malay soldiers conducted patrols, ambushes, and raids against communist strongholds deep in the jungle. Additionally, new technologies were introduced, such as helicopters.

The regiment's commitment to the fight against communism during the Emergency helped maintain stability in Malaya at a time when the region was teetering on the edge of chaos. The regiment was not merely an instrument of British policy but a protector of the Malay people's future. Still, two Scots Guards regiments — a battalion of Grenadiers, and a battalion of Coldstream Guards — were soon brought in to bolster local resources.

Lt-Col Kenneth Molyneux-Carter, CO 2 RMR, brought with him 'a fine offensive spirit and determination to come to grips with terrorists.' It would lead to his being awarded the OBE. His battalion eliminated more terrorists than any other, notching up at least 100 kills along the way, a milestone duly 'celebrated'.

Sergeant Salleh bin Abu Bakar was awarded the Military Medal for his skilful and successful leadership during many months of operations against Communist terrorists. His platoon's achievements were notably attributed to his leadership.

The 3 Battalion was ready to rumble, just nine months after the Emergency was declared. By December 1949 incredibly 4 Battalion was already on jungle patrols.

While there were several engagements resulted in significant casualties — and countless episodes of abject horror and hardship among the local and Commonwealth personnel — one particularly devastating incident involving the Royal Malay Regiment was the Battle of Semur River, near Gua Musang in Kelantan, in 1950.

In this encounter, 12 Platoon of Delta Company 3 RMR was ambushed by MNLA insurgents. The platoon, consisting of 25 soldiers, was conducting a patrol when they came under heavy fire from well-prepared enemy positions along the Semur River.

The initial volley resulted in the immediate death of the platoon commander, 2 Lt Hassan bin Yassin, and several others. Despite being outnumbered and outgunned, the remaining soldiers, under the leadership of Cpl Jamaludin bin Mohamad, mounted a valiant defence. The firefight lasted approximately four hours, during which 17 members of the platoon were killed, and all but two were wounded. The MNLA also suffered significant losses, with 29 insurgents killed during the battle. You might consider it the Emergency's equivalent of the Battle of Bukit Chandu.

**

The British, recognising the importance of fostering a capable military force in the soon-to-be-independent country, continued to invest in the regiment's — and the broader Army's — development.

At that time, 20 Malay officers were attending Sandhurst.

The first-ever colonel was appointed to the regiment: Lt-Gen Hugh Stockwell was former GOC Malaya until tapped on the shoulder by the Queen and Malay rulers to come onboard the Malay Regiment — which he'd shown great interest in— in an honorary capacity, just before he shipped out to Germany. Major Ismail Tahar was now the senior Malay officer in the regiment, and he also had 'welfare' in his duties. The status of Malay Regiment officers was re-evaluated, with an amended bill giving officers of the unit the same status as King's Commissioned Officers. The implication of this was that the regiment would also be liable to serve anywhere in the world, a move welcomed by Capt Hussein bin Onn, who had himself trained with the Indian Military Academy and served in the Middle East in WW2. He felt it would raise the efficiency of the regiment, and that the regiment should be honoured to serve overseas in any case.

Colours were presented to the 1 and 2 Battalions by Sir Henry Gurney, the Federation High Commissioner, urging them to use it as daily inspiration.

'The people of your country look to you to defend them, their families, and their soil against any threat to their security, and to prevent those things happening again that are part of your history,' he told them. 'Let us be quite clear that the danger facing Malaya is militant Communism. Nothing else. When we have finally beaten the Communist attack there will be peace again.'

By December 1951 the 5 Battalion was active in the field.

But things were getting stretched more than a little thin by this rapid expansion. Keeping up with the supply of British officers proved difficult. A temporary solution was found: invite former officers of the late Indian Army to join. Incentives included a recent large pay increase, and former Indian Army officers would be allowed to retain any pensions and compensation they'd already received in leaving their previous posts.

General Sir Gerald Templer, the British High Commissioner to Malaya, was a 'hearts and minds' man, and envisioned a long-term objective for training and leadership establishments in Malaya.

Thus, in July 1952 the Secretary of Defence announced the upgrading of the Malay Regiment Training Depot in Port Dickson to the Malay Regiment Training Centre, which included the establishment of the Pre-Officer Cadet Training Unit and a Boys Company.

'Time has moved on,' noted Raja Lope, 'because in our Boys Company are sons of former pre-war soldiers. They are chips off the old block.'

By December 1952 the 6 Battalion was completely raised, and a patrol near Kuala Lipis in Pahang scored it first 'kill' in its first week out.

The battalions were spread to all corners of Malaya: 1 Battalion at Muar; 2 at Taiping; 3 at Kota Bahru; 4 at Mentakab; 5 at Butterworth; 6 at Raub. A matter of $5,000,000 had been spent on expanding the Port Dickson depot, as well as building permanent HQs for each battalion.

With all this movement came matters of logistics and love. Families did not like to be separated for extended periods. The men naturally wanted their wives and children around, especially as the country became increasingly unsettled and unsafe. But because of the fluid unfolding situation, sometimes it wasn't feasible to build permanent barracks.

This is where Raja Lope — whose son Zamal was then 16 — injected his humanity. When 4 Battalion was posted to Mentakab, and accommodation was ready, he flew 200 wives and children in (safer than by road at that point). The improvement to morale was immediate. And that was something literally close to his heart. He also ensured that men took sufficient leave to remain fresh and fit.

The Royal Malay Regiment alone had accounted for around 245 bandits by this stage.

Operations permitting, Fridays were kept free for men to attend the mosque. In addition all battalions had their own 'Mohameddan Gurus' (an outdated Colonial-era term, these days meaning Islamic Scholar or religious teacher).

Brig Andre had his hands full. Not only with all of this recruitment and expansion, but also with raising a Federation Regiment, Federation Armoured Corps, and pulling all of the units into a cohesive Federation Army.

He was now 53 and due to retire as the first Director of the Federation Military Forces. Several of the FMS rulers asked the War Office if he could stay on, such was the respect for him and his likeable, cheerful nature. Among his final thoughts were that the Boys Company was a 'first-class source' from which to draw future officers of the Malay Regiment and Federation Army. Asked by a reporter what he thought of his men, Andre replied: 'The Malay soldier is stubborn in holding on in the face of desperate conditions. When the need arises he shows a really high standard of bravery and discipline.'

This wasn't just a fanciful response, for this man had seen it for himself side-by-side with his men: 'They proved this when fighting the Japanese. In the Battle of Ulu Semur 25 men proved it again when they faced 250 terrorists and killed 30 of them before they were overwhelmed and most of their own men dead or badly wounded.'

He visited the various units — such as the 1 Battalion in Muar and 5 Battalion in Butterworth — on a round of farewell visits. In Muar, the Corps of Drums was to have played for him on the rugby pitch, but the flooding rains and mists drowned that idea. Instead, they paraded in front of the officers' mess, lit by the headlights of cars!

At his final parade at Port Dickson, the men of the Training Battalion — under Lt-Col C Ford (who passed away within three years of this occasion) — gave an impressive arm-swinging march-past, followed by fifty would-be soldiers of the Boy's Company, who won everyone's hearts that day.

Ford presented *krisses* to Brigadier and Mrs Andre: 'The name of Andre will live in the regimental records for as long as there is a Malay Regiment, for they, more than any other have raised the regiment to its present standard,' he said.

Andre replied: 'I have seen the training of 7000 recruits since the war on the parade ground. They have become efficient soldiers, as we see today. Malaya is proud of her sons.'

Andre was awarded the CBE (Military) the citation being 'for outstanding service to Malaya', and his unsparing devotion to organising the expansion of the Malay Regiment ... all the while being big-hearted and being held in the highest esteem by all ranks. A *Straits Times* headline gushed: 'Man Who *Is* the Malay Regiment.'

He, wife Grace, and six-year-old daughter Margaret then headed to Penang, where they boarded the SS *Corfu*, and sailed to England.

Andre was replaced in the Director of the Federation Military Forces position by his long-time colleague, George Denaro, who had returned from a spell commanding his previous York and Lancaster battalion in Germany.

Within 12 months, the Federation Military College (FMC) was inaugurated in Port Dickson. A flood of 2000 applications washed in for just 100 places, the implication being that Malay youth were very keen to be soldiers.

As the 1950s progressed, the Malay Regiment grew in both numbers and prestige, and Malay officers took on increasingly prominent leadership roles.

In 1952, WW2 veteran Lt Ibrahim Alla Ditta led a patrol under extremely dangerous conditions facing heavy enemy resistance. He demonstrated exceptional leadership, tactical skill, and courage, a significant factor in his receiving the Military Cross.

One of the WW2 veterans, still in service and showing conspicuous bravery, was RSM Ismail bin Baba. His leadership in coordinating troop movements and organising the defence strategies was highly valued in these dangerous conditions. His MBE was a recognition of his dedication and the high regard in which he was held by his fellow soldiers, plus Malay and British officers alike.

Raja Lope, by now a major, was also awarded the MBE. In the past three years, apart from 79 Mentions in Dispatches, the regiment had clocked up an impressive 30 medials and honours, ranging from BEMs and MMs to CBEs and MBEs.

On the occasion of the regiment's 20th anniversary, High Commissioner General Sir Gerald Templer had a message sent to all units operating out in the jungle: 'I hope that your great contribution in the fight against tyranny in Malaya will shortly bring an end to the Emergency so that you and your families may continue to live in peace.'

Just as pre-war, the regiment topped marksmanship competitions; the regiment had not lost its touch. In 1953, it swept the major trophies at the Malaya Rifle Meeting in Kota Tinggi. Capt Zainas bin Isaak of 4 RMR came out as champion individual, while the regiment won the Dunlop Trophy, Sultan of Perak's Challenge Cup, and Lindsay Vears Shield.

By early 1954 — at the height of the Emergency — 21 of the 26 officers in the regiment were Malay, progressively replacing their British counterparts.

At grassroots level, a glowing example of showing skill and courage was 20-year-old Pte Hassan bin Ahmad on patrol with his 'D' Company 1 Battalion platoon. They were near the main road of the Kammunning Estate, Ipoh, when five Chinese were seen acting suspiciously. The patrol challenged the group who immediately ran off, with one of them throwing a grenade, and another pulling out a revolver and firing at the platoon. Pte Hassan killed him with his carbine. That man was identified as Loh Kok, otherwise known as 'The Man with 100 Lives', who'd escaped several earlier scrapes with security forces, always using civilian human shields to make good his escape.

By the following year, Brig Denaro — known widely and affectionately as 'Brigadier George' — had been serving in the federation for 20 years, having joined the Malay Regiment at the time it switched from being an experimental company. Now it was seven battalions large. He praised them as having grown in a short time, especially post-war, into 'quite an efficient force'.

One of the things he pushed for was maintaining the Federal Military College at Port Dickson, which he viewed as 'a mini Sandhurst'. But he also saw value in continuing to send Malay officers to Sandhurst and Eaton Hall. 'They shed their racial identities while in England and come back as true Malayans in form and spirit,' he felt. (Interestingly his son Maj-Gen Arthur Denaro eventually became Commandant of the Royal Military Academy, Sandhurst.)

But for all their history, skill, and training, around 180 British officers died on Malayan soil in the Emergency, many buried at Cheras Road or Terendak Military Cemetery. Things were not as rosy as were being painted to them back home. Extra inducements were offered by the War Office, including interest-free loans on cars and furniture, an extra $58 for those who passed their Malay exam, reduced rents (as low as 7 shillings a week for married quarters) and a bump-up in basic pay by £4 a day for corporals, more for higher ranks. But, those foreigners serving with the Malay Regiment did not get compassionate leave with paid travel back home like their British-unit counterparts.

George Denaro was now 53, and greying at the temples. It was time to retire and return to the UK. He was given a farewell from Kuala Lumpur station (presumably taking the train down to Singapore to board their ship). The High Commissioner Sir Donald MacGillivray was among the well-wishers, as the Malay Regiment band gave them a tuneful send-off.

He was honoured as a Commander of the Most Excellent Order of the British Empire (CBE) for his services to Malaya Command.

And, in a sign of the shifting sands of the times, history was made on 17 August 1956 when 47-year-old Lt-Col Raja Lope — the first Malay to reach that rank, and who'd just become the Raja Kichil Bongsu of Perak, sixth in line to the throne — took the salute as 124 new recruits marched past him at

Port Dickson. *That was him 23 years earlier!* Now he was the first-ever Malay to take the salute.

# Merdeka and Beyond

And so things came full circle in August 1957, when former commanding officer Maj Bruce was invited to ceremonies in Kuala Lumpur celebrating Malayan Independence: *Merdeka.*

Britain was changing, having shed several dominions in Asia over the previous decade. And Malaya was marching towards becoming Malaysia. Bruce had escaped the ill-fortune of being a POW under the Japanese, having been promoted to the rank of Major-General, and appointed GOC Nigeria. He formed, trained, and commanded the 82nd West African Division from 1943 to 1945 — similar to what he'd done so successfully with the Malay Regiment — and led them in operations across West Africa, India, and Burma (where they were part of XV Corps, charged with driving the Japanese out of the colony). He had 28,000 men under his charge.

Two of his first 25 recruits were still in the regiment: Raja Lope (now a colonel) and Mohamed Taib bin Jais (now a Lt-Col). Taib was awarded the MBE for his ongoing efforts, especially as he'd been recently playing the role of an adviser of 'regimental and Malay affairs' as ADC to the Yang di-Pertuan Agong.

Bruce was gratified to learn that the original Chinese tailor Kwong Yuen was still plying his trade, now as the garrison tailor at Port Dickson. Approximately 10-20 per cent of those who fought in the Battle of Malaya — those who had survived the horrific battles, internment, and escaped the traumatic scourge of shell-shock — had continued to serve post-war.

One of these was Pte Tok Ujang, who reported back for duty. (After serving a few more years, he joined the Prisons Department till he retired.)

As the revered 'father of the Malay Regiment', founding officer Bruce was invited to deliver a live radio address on the Malayan Broadcasting Corporation (MBC). In a display of respect and dedication, he painstakingly prepared his speech in Malay, rehearsing it for days. Bruce spoke not of conquest or dominion, but instead of collaboration. He framed the legacy of the Malay Regiment as a testament to what could emerge when two worlds, so often cast as adversaries, found common ground. In his view, the regiment was not merely a relic of colonial authority but a shared endeavour — proof that, at least in this case, the lines between ruler and ruled could blur into something constructive, something lasting.

Most poignantly, he continued on how they'd performed at Pasir Panjang and Bukit Chandu: 'The best news I heard in the war was from a neutral source who described their fighting during the last days before the fall of Singapore.' He concluded with paying respect to those fallen comrades and wishing every success to the regiment and the new country.

**

On the eve of Malaya's independence, Tahir Abdul Majid had the honour of lowering the Union Jack at the Selangor Club Padang in Kuala Lumpur. This symbolic act — performed at the most British of clubs, the social heart of Malayan expat life — accompanied by the British anthem *God Save the Queen*, marked the end of British colonial rule in Malaya. The lowering of the Union Jack was greeted with a loud spontaneous applause and cries of *'Merdeka!'* which reverberated through the city.

The following day, 31 August 1957, the official proclamation of independence took place at Merdeka Stadium. Prince Henry, Duke of Gloucester, represented Queen Elizabeth II. He officially handed over the constitutional instruments to incoming prime minister, Tunku Abdul Rahman, marking the formal transfer of sovereignty. Sir Donald MacGillivray, the last British High Commissioner in Malaya, also attended the ceremony.

The new federation flag was raised, and Tunku Abdul Rahman led the crowd in chanting *'Merdeka!'* seven times, a spiritually significant number in Islamic culture.

And so a new country was born: The Federation of Malaya. And a new flag was flown across the country. Its 11 red and white stripes represent the states and the federal government. A blue canton signifies harmony among its many peoples, housing a yellow crescent and star (yellow being the colour of royalty, their constitutional monarchy). The crescent represents Islam, the country's official religion, while the 11-pointed star symbolises unity among the states and the federal territories. Their own flag to fight under.

**

By this time, the Malay Regiment had transformed into a professional and modern military force, ready to defend the sovereignty of the newly-formed nation. It was a significant passing of the torch from British colonial rule to self- governance, with the regiment standing tall as both a symbol and protector of the new Federation of Malaya.

The regiment's evolution during this period was not just about military capability; it was about nation-building. As the independent nation took its first baby steps, the Malay Regiment was there, holding its hand and ready to guard its newfound freedom.

But of course, the Emergency would continue for three years more after the Federation flag was raised in Merdeka Square. (The Communists straggled on, with Chin Peng living in exile in southern Thailand, and a peace accord was only finally signed in 1989.)

**

1958 was a watershed year. After World War II, Abdul Manan bin Abas — the only surviving Malay officer from Bukit Chandu in 1942 — continued his military career, and achieved the rank of Lieutenant-Colonel. He became the first Malay Commanding Officer of 4 RMR. He was awarded the Military Cross for his exceptional bravery and leadership during the Malayan Emergency. The same year, Lt-Col Tun Ibrahim Ismail was appointed as the first Malay Commanding Officer of 6 RMR. In a distinguished career, he would go on to become a *Jeneral.*

These two appointments marked a significant transition from British to local leadership in the Malaysian military.

The last British officers that served with the regiment are not readily recalled by anyone these days. Among them were Capt E Lethall and RSM Harold Foster.

Capt E Lethall was the Director of Music. As bandmaster with the Central Band, his creation of the regiment's music traditions earned him a last-minute pre-Merdeka accolade: an MBE. Foster (formerly of the Gordon Highlanders) had also earned an MBE for his contributions towards the efficiency of the regiment, having trained thousands of recruits in military discipline and drill. The Federation Armed Forces put out a statement when he retired in July 1958 after six years with the regiment: 'Undoubtedly, much of the reputation of the Royal Malay Regiment for smartness and bearing on parade is due to the teaching and example of RSM Foster.'

But Capt CF West, who appears to have taken over from Lethall as Director of Music, was the only European on parade in March 1959, when Lt-Col Khalid bin Hashim took the salute at a passing out parade in Port Dickson.

In the face of Malaysianisation, the participation of foreign officers — just as in many industry sectors — essentially fizzled out. The regiment could now stand on its own two feet.

**

Then came their first overseas deployment: Congo, Africa. From 1960 to 1963, the Royal Malay Regiment participated in the United Nations Operation in the Congo (ONUC), contributing to peacekeeping efforts during the Congo Crisis. Tasked with maintaining order and stability, they conducted patrols, protected civilians, and guarded strategic installations amid violent conflict between local factions.

April 1961 saw emotional scenes as the second batch was farewelled from 5 RMR camp in Ipoh Road, KL. On the Padang were 80 men plus two officers — dressed in full jungle greens with kit — headed to the Congo. Receiving their salute was Deputy GOC of the Federation Army one Brig Raja Lope Nor Rashid. From a private, he'd risen to be the most senior Malayan officer in the army. He was 52 and was retiring the very next day. He told the men that Malayan soldiers had attained a high standard locally and internationally and must do their bit to keep it so. 'I have enjoyed every moment of my career as a soldier. I am sad to go.' They were losing a real champion in every sense.

Three cheers broke out for the brigadier.

But that day was more about the job that lay ahead for these warriors departing to Africa. Haji Dusin bin Ugama was the chief religious instructor for the Armed Forces and he said prayers for their safe journey. And what a journey it would be: by truck to Rawang, by train to Butterworth, by Royal Air Force plane to the Congo.

Raja Lope, with his easy common touch, shook hands with the men as they climbed onto their trucks. The RMR band struck up *Auld Lang Syne*, and then it was time for tears: girlfriends, wives, mothers alike gave teary embraces to their men.

'Goodbye. Pray for us,' shouted one of the men as the trucks began to roll out. More tears.

Looking on were a further 618 men who would be likewise leaving in the next week or so, making it a full battalion-sized contingent of 1400 men under Brig Abdul Hamid bin Bidin.

The challenges for the Malayan Special Force would have been vast: first exposure to a whole new continent; local languages and cultures that were unrecognisable; and even the *matsellehs* were different because they were mainly Belgian. Yet, the regiment earned international recognition for their professionalism and discipline under challenging conditions.

**

While the Congo was in meltdown, back home peace didn't last long either. The Malay Regiment found itself almost immediately back in action again for the Indonesia-Malaysia *Konfrontasi* from 1963-5. *Konfrontasi* was a conflict initiated by Indonesia under President Sukarno in opposition to the formation of Malaysia, which combined Malaya, Sabah, Sarawak, and Singapore.

The new-improved Malaysia was born in 1963, with the inclusion of those latter three states in the federation (although Singapore would itself declare independence in 1965). The flag was adapted to reflect this addition, with 14 stripes, and a 14-pointed star. (Over the years, the Malaysian flag would come to be known as the *Jalur Gemilang*, the 'Stripes of Glory').

Sukarno viewed this expanded federation as a neo-colonial plot, fearing British influence and the potential disruption of regional power dynamics. The conflict saw Indonesian forces conducting raids, sabotage, and incursions into Malaysian territories, particularly in Borneo. The Malay Regiment conducted operations along the border areas of Sabah and Sarawak, engaging in patrols, ambushes, and counter-insurgency missions to thwart Indonesian infiltrations. Their intimate knowledge of the local terrain and cultural landscape was instrumental in maintaining security and stability during this period.

Once again, the Royal Malay Regiment was in the thick of the action in those three years, but perhaps no incident more notable than the Kalabakan Incident, December 1963.

Two months earlier, Indonesia deployed the *Korps Komando Operasi (KKO)*, an elite marine unit, to Sebatik Island near the Sabah border. They established training camps and prepared for incursions into Malaysian territory. Tensions escalated with several border skirmishes and, in early December, an Indonesian Tupolev Tu-16 bomber flew over Tawau Bay, bombing the town twice.

Three weeks later, a force of 35 KKO regulars and 128 volunteers crossed into Sabah. Their mission was to capture Kalabakan and then advance towards Tawau, anticipating support from Indonesian expatriates in the area.

At approximately 11pm on 29 December, they launched a surprise attack on a position held by 'C' Company of 3 RMR, commanded by Major Zainal Abidin bin Haji Yaacob. The Malaysian troops were caught off guard, resulting in eight soldiers killed, including the major, and nineteen wounded. Sgt Daud bin Ninggal, despite being wounded, displayed exceptional bravery in engaging the enemy, his courage earning him a first Military Medal.

In response to the attack, Gurkha reinforcements were flown into the area. Over the next month, combined Malaysian and Commonwealth forces conducted operations to track down the infiltrators. Approximately two-thirds of the KKO participants were killed or captured. Captured insurgents revealed that they had expected the local population to rise and support them as liberators, a belief that was based more on hopefulness than reality.

The 3 Battalion was not short of heroes. Sergeant Mohamed Nor bin Hassan was recognised for his exceptional bravery and leadership during jungle operations. He consistently demonstrated himself to be an aggressive, skilful, and fearless leader on numerous occasions. His actions significantly contributed to the successes achieved by his platoon against Communist terrorists, hence winning the Military Medal, same as Sergeant Hussein bin Pak Su.

The Kalabakan Incident marked a significant escalation in the Indonesia–Malaysia Confrontation, highlighting the volatility of the region for the next couple of years.

Sgt Daud's second Military Medal was awarded for his valour during an operation in Sarawak where he led his section in a successful assault against Indonesian forces, demonstrating outstanding leadership and courage under fire.

Tensions eased after Sukarno's fall in 1965, and the conflict officially ended with a peace agreement in August 1966 under President Suharto's administration.

By now the regiment boasted eight battalions, with talk of raising more, supplemented by opening up recruitment to indigenous people in Sabah and Sarawak, who had proved so useful to the Allies in WW2 as trackers, and generally had an innate understanding of the terrain. They were seen as a natural fit. Some were already part of the recently-formed Royal Ranger Regiment.

**

Maj-Gen Bruce died on 6 February 1966, at Queen Alexandra's Military Hospital in Westminster. At his bedside, holding his hand, were his second wife Mona and her daughter Lavinia. The cause of his death was cerebral haemorrhage, brought on by hypertension. He was 69. In a touching gesture to their revered founding officer, the Chief of the Malaysian Armed Forces General Tunku Osman — who had served with British units including Force 136 during the war, before joining the Malay Regiment in 1947 — ordered all units of the Royal Malay Regiment to fly their flags at half-mast.

Port Dickson remains the regiment's HQ and spiritual home, while other specialist battalions, such as the paratroopers of 9 Battalion, use the Terendak Camp in Malacca.

**

In 1989–90, the regiment contributed to the United Nations Transition Assistance Group (UNTAG) in Namibia, supporting the nation's transition to independence from South African administration. The RMR's responsibilities included monitoring ceasefires, overseeing the repatriation of refugees, and ensuring the withdrawal of foreign military forces. Their presence was vital in maintaining peace and stability during the electoral process, facilitating Namibia's own journey to sovereignty.

Next up in its Battle Honours, from 1992-93, the regiment participated in the United Nations Transitional Authority in Cambodia (UNTAC), contributing to peacekeeping efforts during Cambodia's transition to democracy. The RMR's duties included maintaining security, overseeing ceasefire agreements, and supporting the organisation of free elections. Then, back-to-back, they served in Somalia as part of the United Nations Operation in Somalia (UNOSOM II), aimed at restoring order and providing humanitarian aid amid the country's civil war. In the notorious battle of Mogadishu in 1993 — as depicted in *Black Hawk Down* — the RMR carried out critical tasks, including escorting aid convoys, protecting UN personnel, and stabilising conflict zones. They navigated this extremely hostile environment, contributing to Malaysia's growing reputation as a committed and reliable partner in international peacekeeping operations.

The Royal Malay Regiment then participated in the United Nations Protection Force (UNPROFOR) and subsequent peacekeeping missions in Bosnia and Herzegovina during the Balkan conflicts. Under the command of Major-General Md Hashim Hussein in 1993-4, the RMR was instrumental in maintaining ceasefires, protecting humanitarian aid convoys, and ensuring the safety of civilians in conflict zones. Their professionalism contributed to the stabilisation efforts in the region. Overall, this tied them up for five years till 1998.

It was quite a while then before they were engaged in Operation Daulat, a military operation conducted by the Malaysian Armed Forces and Royal Malaysia Police in response to the Lahad Datu standoff in Sabah, Eastern Malaysia, in February/March 2013. There, a group of about 200 armed men claiming to be the 'Royal Army of the Sultanate of Sulu' from the Philippines landed in Lahad Datu. They sought to reassert a historical claim to Sabah, which they considered part of the Sulu Sultanate. The RMR's role was to eliminate the armed group and restore security to the region. They conducted combat operations, area sweeps, and secured key locations alongside elite units like the *Gerak Khas Kommandos* (GGK) and VAT 69 Commando police units. Within a month, the Malaysian forces had successfully defeated the intruders after intense skirmishes, reaffirming Malaysia's sovereignty over Sabah.

Apart from regional war game exercises, this was the Royal Malay Regiment's most recent battle engagement at the time of writing.

Despite its modernisation, the Royal Malay Regiment remains deeply connected to its historical roots and traditions. Ceremonial duties, such as guarding the royal palaces and participating in state events, are carried out by elite units of the Royal Malay Regiment. These ceremonial responsibilities are a reminder of the regiment's role as not only a military force but also a symbol of national unity and continuity. The regiment's deep respect for tradition is also reflected in its continued loyalty to the Malaysian monarchy. The Sultan of each Malaysian state holds a special position as ceremonial head of the regiment, and this connection to the royal families underscores the regiment's cultural and historical importance within the broader fabric of Malaysian society.

'Look after the country we defended with our blood and our lives, do not destroy it,' said Pte Ujang 'Tok' Mormin — the last surviving member of the the Malay Regiment who fought in World War Two — shortly before he died at the age of 100 in 2021. He was fittingly given a military funeral and buried in the Royal Malay Regiment cemetery in Port Dickson.

Guided by the homegrown heroic spirit epitomised in the battle of Bukit Chandu, the regiment's story is ultimately one of continuity, resilience, and national pride. The experiment proved a success. Mission accomplished: Tick. Malaysia *boleh!*

—The End—

If you have enjoyed this story, please take two minutes to post a review on Amazon, Goodreads or Facebook, etc, so that other like-minded readers might know about the amazing deeds of the Malay Regiment, too.
Thanks in advance.

# Awards to the Malay Regiment for service in World War II

**Distinguished Service Order:**

Major James Richard Glencoe Andre (17888), The Lincolnshire Regiment.

Major George Tancred Denaro, (27507), The York & Lancaster Regiment.

**Military Cross:**

Captain (acting Major) Griffith Powell Richards, (41421), The South Staffordshire Regiment (attached the Malay Regiment).

Lieutenant Alistair George Mackenzie (225856), General List (attached The Malay Regiment).

2nd Lieutenant (acting Captain) Edward Howard Stanley Bretherton (26935), General List (attached The Malay Regiment).

**Military Medal:**

Private Yaacob bin Bidin (182)

**Mention in Despatches:**

Lieutenant-Colonel John Talbot Bretherton Hawkshead-Talbot MC (8257).

Lieutenant Abbas bin Mohd Saiid (1780).

Lieutenant Adnan Saidi (90).

Lieutenant TPD Jones (204896).

2nd Lieutenant Abbas bin Abdul Manan (283).

2nd Lieutenant Noor Din bin Ibrahim (315).

Sergeant Ali Hashim bin Hashim (63).

Corporal Abdul Rani bin May Hussein (246).

L/Cpl Jamal bin Haji Chin

Private Abdul Wahab bin Mohamed Taib (2330).

# Bibliography

*A Bleeding Slaughterhouse: The Outrageous True Story of the Alexandra Hospital Massacres, Singapore*, February 1942. Stuart Lloyd.

*A Most Splendid Soldier: The Life of My Great-Grandfather Major General George Bruce,* Orson Fry, 2017, Bard Senior Project.

*Guns of February: Ordinary Japanese Soldiers' Views of the Malayan Campaign in 1941*, Henry Frei.

*History of the Malay Regiment 1933-1942, Journal of the Malaysian Branch of the Asiatic Society,* Vol 38 Pt 1, Dol Ramli, 1965.

**Other Sources**

*A Good Soldier Obeys Orders, Pinang Gazette and Straits Chronicle,* 9 February 1935.

*British Rule Vindicated. Malaya's Great Welcome,* The Times, September 25, 1945.

drshafie.blogspot.com

*Far East Conflict Still Powder Magazine, Pinang Gazette and Straits Chronicle,* 10 June 1938.

*Great Future for Malay Regiment Predicted: GOC-in-C's Farewell Message, Indian Daily Mail,* 9 October 1946.

KaisersCross.com

*Malaya's Finest Regiment, Alan The Leam Seng, New Straits Times,* 21 June 2022.

*Malay Regiment's Fine Display, Straits Times,* 29 November 1934.

*Malay Regiment Says Good-bye to Brigadier Andre, Straits Times,* 17 November 1952.

*Malay Regiment Strength, Straits Times,* 19 December 1946.

*Malay Regiment To Be Reformed, Indian Daily Mail,* 6 March 1946.

*Man Who Is The Malay Regiment, Straits Times,* 17 November 1952.

Memoirs of Douglas Morris far-eastern-heroes.org.uk

NewspapersSG eresources.nlb.gov.sg

Portal Rasmi Majlis Daerah Bera (MDB) mdbera.gov.my

*Private Ujang Mormin, last survivor of Battle of Pasir Panjang during WWII, dies due to Covid-19, Straits Times,* 10 February, 2021.

Reflections at Bukit Chandu exhibits, Singapore.

*Remembering the last veteran of the Battle of Pasir Panjang,* 12 February 2021, FreeMalaysiaToday.com

roll-of-honour.org

*Royal Major Looks After Malay Soldiers' Welfare, Straits Times,* 17 May 1953.

Royal Malay Regiment, militarybadgecollection.com

*Senshi Soshi battle report,* 15 February 1942

*Shropshire family's epic wartime escape,* 15 July 2020, Shropshirestar.com

*The Malay Regiment: Abbas Abdul Manan's oral history interview in 1992,* Youtube

Reflections at Bukit Chandu
nhb.gov.sg/bukitchandu

# Photograph Copyrights

**Front Cover:**

Men of the Malay Regiment at bayonet practice, on Singapore Island, 1941: Palmer (Lt), War Office official photographer. Public domain.

Portrait of Adnan Saidi: photographer unknown.

Lt-Col Bruce portrait: photographer unknown, copyright Orson Fry.

Pboto page 1: Lt-Col Bruce portrait, wedding photo, and house: copyright Orson Fry.

Page 2: Insignia: militarybadgecollection.com. Songkok: Malay Heritage Centre.

Page 3: 1933 and 1934 group photographs: photographer unknown.

Page 4: Port Dickson aerial photo and card: courtesy Paul Pilkington.

Page 5: Malay Regiment, Commissioned Officers receiving their swords, Kuala Lumpur; 4.11.1936. Arkib Negara Malaysia Accession No. 20010022875

Exham and Saidi: still from Abbas Abdul Manan interview video, YouTube.

Page 6: Stills from Five Faces, produced by Malayan Colonial Government 1938.

Page 7: Malay Regiment rifleman: photographer unknown.

Page 8: Bayonet practice: WikiCommons License.

Pillbox photograph: Stuart Lloyd.

Page 9: Japanese attacking Opium Hill, Australian War Memorial. AWM 127901.

Page 10: Portrait of Adnan Saidi: photographer unknown.

Page 11: POW card: source rollofhonour.org

Howard Bretherton award: courtesy Merryn Godber.

Page 12: Raja Lope: sembangkuala.wordpress.com. Source: Tengku Yasmin Tengku Yusof.

Adnan Saidi bust: photograph Stuart Lloyd.

**Back Cover:**

1934 group photograph: photographer unknown.

Mortar crew: still from Abbas Abdul Manan interview video, YouTube.

Reasonable attempts have been made to identify and contact copyright holders of text and photographs used in this publication. If crediting of your work has been inadvertently overlooked or omitted please contact the publisher and we will make amendments if/where required in the next edition.

# Other war books by Stuart Lloyd you might find interesting

*A Bleeding Slaughterhouse: The Outrageous True Story of the Alexandra Hospital Massacres, Singapore, February 1942.*

*The Missing Years: A POW's Memoir from Changi to Hellfire Pass.* (Also in audiobook.)

*The Grubby Little Men Who Raped Hong Kong: The True Story of the St Stephen's College Massacre, December 1941.* (Also in audiobook.)

*The Depths of December: The Sinking of HMS Repulse, Prince of Wales ... and the British Empire.* (Also in audiobook.)

**Other books by Stuart Lloyd you might find enjoyable**

*Tales from the Tiger's Den: An Oral History of Foreigners in the Far East 1920-2020.*

*Gone Troppo! My search for Tropical Paradise.*

*Honeymoon for One: Collected travel writings from Australia to Zimbabwe (and everywhere in between).*

*Bamboozled: The Lighter Side of Expat Life in Asia.*

*Hare of the Dog: History, Humour, and Hell-raising from the Hash House Harriers.*

# About the Author

A sixth-generation Southern African, Stuart was born in Rhodesia (Zimbabwe) before emigrating to South Africa then Australia. In Asia he has lived in Hong Kong, Singapore, China, and Thailand, for nearly 25 years, and has travelled to 75 countries worldwide.

He has conducted special-interest military history tours through Singapore, Malaysia, and Thailand, and conducted WW2-related research in Hong Kong, Singapore, Philippines, Thailand, Burma, Papua New Guinea, and Malaysia.

Stuart was once described by *The Telegraph*, UK, as 'The perfect storyteller.'

**For more visit stuartlloyd.net**

Subscribe to Stuart's 'Emotional Journeys' blog for sneak previews, additional content, and more inspiring human spirit stories.

**Connect:**

**Instagram: @RealStuLloyd**

**Facebook: @StuLloydStorytelling**

Made in the USA
Monee, IL
04 November 2026

67385352R00059